THE DIPPERS

Frontispiece

The five species of dipper
Top right   White-throated Dipper *Cinclus cinclus gularis* – Britain.
Top middle   *C. c. leucogaster*, the white-bellied race from E. Russia.
Top left   American Dipper *C. mexicanus*.
Centre   Brown Dipper *C. pallasii*.
Bottom left   Rufous-throated Dipper *C. schulzi*.
Bottom middle and bottom right   Two races of White-capped Dipper – *C. leucocephalus leuconotus* (middle) and *C. l. leucocephalus* (right).

THE DIPPERS

STEPHANIE J. TYLER
and
STEPHEN J. ORMEROD

With illustrations by
DARREN REES

T & A D POYSER
London

First published 1994 by T & AD Poyser Ltd
Print-on-demand and digital editions published 2010 by T & AD Poyser,
an imprint of A&C Black Publishers Ltd, 36 Soho Square, London W1D 3QY

www.acblack.com

ISBN (print) 978-1-4081-3681-2
ISBN (epub) 978-1-4081-3680-5
ISBN (e-pdf) 978-1-4081-3679-9

A CIP catalogue record for this book is available from the British Library

This is a print-on-demand edition produced from an original copy.

It is produced using paper that is made from wood grown in managed sustainable forests. It is natural, renewable and recyclable. The logging and manufacturing processes conform to the environmental regulations of the country of origin.

Printed and bound in Great Britain

MIX
Paper from responsible sources
FSC
www.fsc.org
FSC® C013604

Contents

List of Photographs

List of Vignettes

List of Figures

List of Tables

Preface

As a child living in Lincolnshire I had no opportunities to see Dippers. I vividly remember my excitement though when, as a 10-year-old, on a family holiday in Scotland, I finally saw my first Dipper on the shore of Loch Tay. Other sightings followed on subsequent holidays to the Lake District and the west country.

My more recent interest in Dippers is entirely fortuitous and due to a long-standing infatuation with Grey Wagtails which started when I was doing some fieldwork on the feral ponies in the New Forest back in the late 1960s. After moving to Wiltshire, where I spent much time along rivers in search of Grey Wagtails and their nests, I came into regular contact with Dippers on the By-Brook in the north of the county. A long period in the Middle East and East Africa then intervened, but eventually a return to the UK, and a home in the Wye Valley with Dippers virtually on my doorstep, led me to a more serious study of these remarkable birds from 1978.

In 1983 I met Steve Ormerod, who was based at the University of Wales' field study centre at Newbridge on Wye in mid-Wales. As a freshwater biologist he had become interested in Dippers through their aquatic prey. Close collaboration over research on both Dippers and Grey Wagtails in mid- and south-east Wales ensued, resulting in our gathering much new data on the ecology of these river birds, and highlighting the adverse impact on Dippers of acidification of so many upland streams in Wales. We have visited up to 200 nests and about 100 roost sites each year, and have ringed well over 7000 Dippers in mid- and south-east Wales since 1978, now having retrap histories on well over 1000 of these. This fieldwork has provided us with much information on breeding patterns, on movements, mortality and fidelity to both nest and roost sites and to mates. It has also given us countless hours of pleasure walking along some of the finest rivers and streams in Britain.

Our fascination with the whole family of dippers has taken us from Scandinavia to Morocco to look at other populations of the Eurasian or White-throated Dippers, all slight variations of the species to which our Welsh Dippers belong. We have journeyed, too, up into the Himalayas for Brown Dippers, to the Rockies and the Sierra Nevada for the American Dipper and all too briefly to Ecuador for the delightful White-capped Dipper. Finally, in October 1993 I went with my husband to north-west Argentina in search of the rarest and most restricted species of dipper, the endangered Rufous-throated Dipper, and we were fortunate not only to see and photograph this small dipper but also to find some occupied nests.

Dippers whether in Europe, Asia or the Americas, occur in exciting habitats, amongst those least modified by man — the fast-flowing mountain streams, rocky, wooded gorges and waterfalls. Sadly even these all too often now fall victim to huge dam projects, hydro-schemes, reservoirs, irrigation projects or pollution from domestic and industrial effluent. We hope that throughout the ranges of the five species of dippers there will always be untamed torrential, rushing rivers where the remarkable behaviour of this unique family of passerines can be observed. In the words of the observant if somewhat anthropomorphic naturalist, John Muir, who was a great champion of wilderness areas:

> 'Find a fall or cascade, or rushing rapid, anywhere upon a clear stream, and there you will surely find its complementary ouzel, flitting about in the spring, dining in the foaming eddies, whirling like a leaf among the foam-bells; ever vigorous and enthusiastic, yet self-contained, and neither seeking nor shunning your company . . . He is the mountain stream's own darling, the hummingbird of blooming waters, loving rocky ripple slopes and sheets of foam as a bee loves flowers, and a lark loves sunshine and meadows.'

This book mainly refers to the White-throated Dipper in Britain, drawing extensively on our own work. We have, however, tried to do justice to the results of all the work on this species carried out in Europe and further afield, and that on other species of dipper. Hence we frequently compare our results with those of researchers in Europe, and where relevant we refer to information on the three New World species and on the Brown Dipper. Throughout the book we use **D**ippers to mean Eurasian or White-throated Dippers, and **d**ippers when referring generally to members of the family Cinclidae.

Our enthusiasm for dippers is shared by many other observers with a resulting extensive literature on the family, particularly in Germany. Gerhard Creutz, for example, has written much on 'Die Wasseramsel', with a monograph in 1966. The December 1985 issue of *Okologie der Vögel* was devoted exclusively to dippers, as was Volume 31 of *Egretta*, Vogelkundliche Nachrichten aus Osterreich. Dippers have also attracted the attention of amongst others, Balat in the former Czechoslovakia, and various Scandinavian, Spanish and Russian researchers. In 1991, Volume 4 of the journal *Cinclus Scandinavicus* covered many aspects of the Dipper's biology in Norway, Sweden and Finland. Russian and Japanese workers have written on Brown Dippers, and Bent, Bakus, and Price and Bock in particular on the North American species. Unfortunately the South American species have been little observed, although in the mid-1980s Sergio Lasso made a brief study of White-capped Dippers near Quito in Ecuador. What little was known of the rare Rufous-throated Dipper up to our brief visit in 1993 is well summarised by Neils Krabbe in *Threatened Birds of the Americas, The ICBP/IUCN Red Data Book* (Collar *et al.*, 1992).

Acknowledgements

Many other people have contributed to our Welsh studies. In particular we thank Jerry Lewis for so readily sharing his data from the Grwyne Fawr and Grwyne Fechan in Gwent and Breconshire with us, and Neil Allinson, Tony Cross, Dominic Hudson, Sally Moralee (née Pester), Dr John O'Halloran and Dr Sonya Yoerg who all worked for two or more years on the Welsh Dipper research; we are immensely grateful to these observers and to Andrew Lucas, Sarah Cartmel, Mick Green and Richard Jenkins who were also contracted for various periods. Volunteers made important contributions too. Amongst these we would like to mention Mark Boilstone, Felicity Burge, John Flynn, John Icke, Richard Lucas, Jon Stratford and particularly Lindsay, Robert and Sally Tyler.

Our autumn Dipper weekends from 1986 to 1993, when volunteers helped trap birds at roost and in nets as part of an annual monitoring exercise, have been well supported and we sincerely thank all participants. They are too numerous to mention individually, but we single out for special thanks the more regular attenders for their hard work: Pamela Allen, Dave Bull, Graham Couchman, Dave Hanford, John High, Rowena and Reg Langston, Jill Moss, Bob Sugden, Dr Juliet Vickery, Jeremy Wilson and their EGI teams and Heather Woodland, and especially Tony John, Dr Dorian Moss, Gwyn Roberts and Rupert and Margaret Wilson who helped on most if not all of the weekends, and Dr Fred Slater for facilities at the Llysdinam Field Centre at Newbridge on Wye.

Many other dipper researchers have freely made their data available to us for which we are most grateful. These include Dr David Bryant, Svein Efteland, Dr Rupert Horvath, Sergio Lasso, Derek Lees-Smith, Dr David Parkin, Ken Perry, Jesus Santamarina, Maurizio Sara and Gabriele Sorci and Geoff Shaw, as well as BTO staff. Funding for various aspects of our work came from the former Central Electricity Generating Board (CEGB), the former Welsh Water Authority, the National Rivers Authority, the World Wide Fund for Nature, the British Ecological Society, Rechem International and the Royal Society for the Protection of Birds (RSPB). We are grateful especially to Dr Colin Bibby, formerly head of research at the RSPB, and James Cadbury (RSPB), to Professor Ron Edwards and to Dr Dave Brown and Dr Bob Aston (both formerly of the CEGB) for enabling funding.

We owe a great debt to Mike G. Wilson of the EGI at Oxford, and to Barbara Scott and Ros Tudor (née Ottoway) for help with translations of foreign references, and to Lynn Giddings and Ian Dawson of the RSPB library for cheerfully finding numerous references for us. Other people whom we wish to thank include Margaret Anthoney and Sheila Poulton of the RSPB for their infinite patience over typing tables; Richard Farmer, Richard Porter, David Pritchard and Kevin Standring of the RSPB and Bernd Mlody; Dr Jim Fowler, Dr Malcolm Greenwood, Dr John A'Brooke and J.E. Chainey of the British Museum for their help over ectoparasites on Dippers and in their nests, Dr Hugh Jones for looking at blood samples for blood parasites, Dr John A'Brooke and Martin Townsend for help with faecal analysis, Dr Keith Bull, Michael French and other staff at the Institute of Terrestrial Ecology at Monkswood for analysis of contaminants in Dipper eggs, and Dr Neils Krabbe and Sergio Lasso for contributing information on Rufous-throated and White-capped Dippers.

Dipper in classic habitat

CHAPTER 1

Introduction: Dippers, their origins and taxonomy, distribution and status

Gigantic wrens or strange thrushes? What are dippers' closest relatives? Where did this unique family evolve and how many species of dipper now exist?

In this chapter we discuss the origins of dippers (Cinclidae), a family of quite remarkable aquatic passerines. We describe the various species of dipper, their distribution and status, and differences between geographical populations of some of these, notably of the most widespread species, the Eurasian or White-throated (formerly White-breasted) Dipper *Cinclus cinclus* (hereafter generally referred to as the Dipper).

DIPPER NAMES AND THEIR DERIVATION

Cinclus, the generic name of dippers, literally means a bird that has the habit of moving its tail. Dippers frequently raise and lower their tails during courtship displays, and the cocked tail is characteristic of the family. The similar generic name, *Cinclodes*, refers to a family of South American birds, sometimes called shaketails, that live along rivers, and similarly move their much longer tails.

The common name, Dipper, clearly refers to the birds' strange bobbing movement. This name is now widely used throughout Britain, but in the past there were numerous local names for this bird (see Swainson, 1886). Some of these refer to the bird's black and white or pied plumage; hence such names as 'piet' or 'pyot', 'water pyot' and 'river pie'. Other local names include 'ess cock' (Aberdeen), 'water peggie' (Dumfries), 'water colly' (Somerset), 'water crake', 'water thrush' (Cornwall) and 'Bessie Ducker'.

Water Ousel was commonly used and in parts of north-west England, as in the Longendale area, the Dipper was referred to as a Rock Ousel (Coward and Oldham, 1910). This, confusingly, was also the name used for Ring Ousel on the Yorkshire/Lancashire border.

In Scotland, Watson (1972) notes the names 'burnie baker' which refers to the bird's white 'apron', and 'water blackbird' or 'water craw'. The latter names were also used in the north of England and Ireland. In these areas as well as in parts of Scotland, Dippers were once believed to be female Kingfishers, male Kingfishers alone being blue and orange. Perhaps the Dipper's fast and direct flight, akin to that of a Kingfisher, may have given rise to this belief.

The Gaelic names 'Gobha nisge' and 'Gobha dhubh nan alt' translate as Water Blacksmith and Blacksmith of the stream, or 'bird from the black water'. 'Mwyalchen y dwfr' is an ancient British name (Forrest, 1907). In Wales, there were also numerous local names for the Dipper. Forrest (1907) mentions some of these: 'Mwyalchen ddwr' and 'Aderyn du'r dwr' literally translate as Water Blackbird, 'Bronwen y dwr' as Water Whitebreast and 'Bronwen y garw' as Whitebreast of the torrent. 'Trochwr' and 'merwys' mean Plunger and Dipper, whilst 'wil y Dwr' and 'Iar Ddwr' are Water Willy and Waterhen. In Co. Mayo in Southern Ireland Perry (1986) found farmers who knew of the dipper as a 'wee waterhen' or by its Gaelic name Gabha Dubh.

The American Dipper was also known as the water ousel by John Muir and earlier writers. Muir also referred to it in the Sierras of California as the water thrush, whilst in Montana, Bakus (1959a) heard it called 'teeter-bird'. This is misleading because the dipper does not actually teeter or wag its tail up and down like the Spotted Sandpiper which is better known as the teeter bird or 'teeter-arse'.

In other countries of Europe the Dipper has inspired many evocative names. For example, in Finland it is called 'pick-axe of the rapids' (Koskikara) whilst the Polish 'pluczsz' literally means 'splash' referring to the sound a Dipper makes when diving into water. Elsewhere it has more prosaic names. Thus in France names include Cincle plongeur and Merle d'eau (water blackbird); in Spain, Germany and Holland too it is described as a 'water blackbird' (Mirlo acuatica, Wasseramsel and Waterspreeuw), whilst the Swedish Strömsare literally means a 'stream starling'.

THE TAXONOMY AND ORIGINS OF DIPPERS

Most taxonomists now accept that there are five species of dipper, although some (e.g. Campbell and Lack, 1985) still consider the two South American species to be the same species. We follow Collar *et al.* (1992) in separating these two Andean dippers.

All five species are remarkably similar in general appearance, behaviour and nesting habits, and all inhabit fast-flowing upland streams in different parts of the world (Fig. 1). They differ only in size, plumage and one or two behavioural traits.

The White-throated Dipper (the Dipper throughout this book) is the familiar dipper of Britain and Europe; it occurs too in North Africa and across Russia to the Himalayas and China. Overlapping with this widespread species in Asia is the large uniform-coloured Brown Dipper *C. pallasii*, which extends across into Japan and up into China.

A third species occurs down the mountain chain of western North America from the Canadian Rockies to the Sierras of California and down through Mexico to Costa Rica and Guatemala. This American Dipper *C. mexicanus* is similar to the Brown Dipper in its uniform plumage but is smaller and greyer. In the rivers of the Andes in South America lives the striking White-capped Dipper *C. leucocephalus*, whilst further south in a narrow strip of forest in south-west Bolivia and north-west Argentina lives the rarest of all five species, the little-known Rufous-throated Dipper *C. schulzi*. This is grey-brown with a rufous throat and, like the White-capped Dipper, has white on the inner edges of the primary feathers of the wing. These white edges are more extensive in the Rufous-throated Dipper than in the White-capped Dipper, extending on to the inner webs of the secondaries in the former species. Both these South American species exhibit wing-flicking behaviour, when the white wing patch is revealed, as well as or instead of the usual bobbing of the body or dipping. They rarely cock their short tails although at each wing-flick the tail is depressed.

Discussions have ranged over the years about the group of birds from which dippers evolved and about the area in which they originated. It was variously suggested that dippers evolved from wagtails (family Motacillidae), from the North American water-thrushes or even from the Timaliidae, a mainly African family including the babblers and chatterers. By the late 19th century, most ornithologists were inclined to think that wrens (Troglodytidae) or thrushes

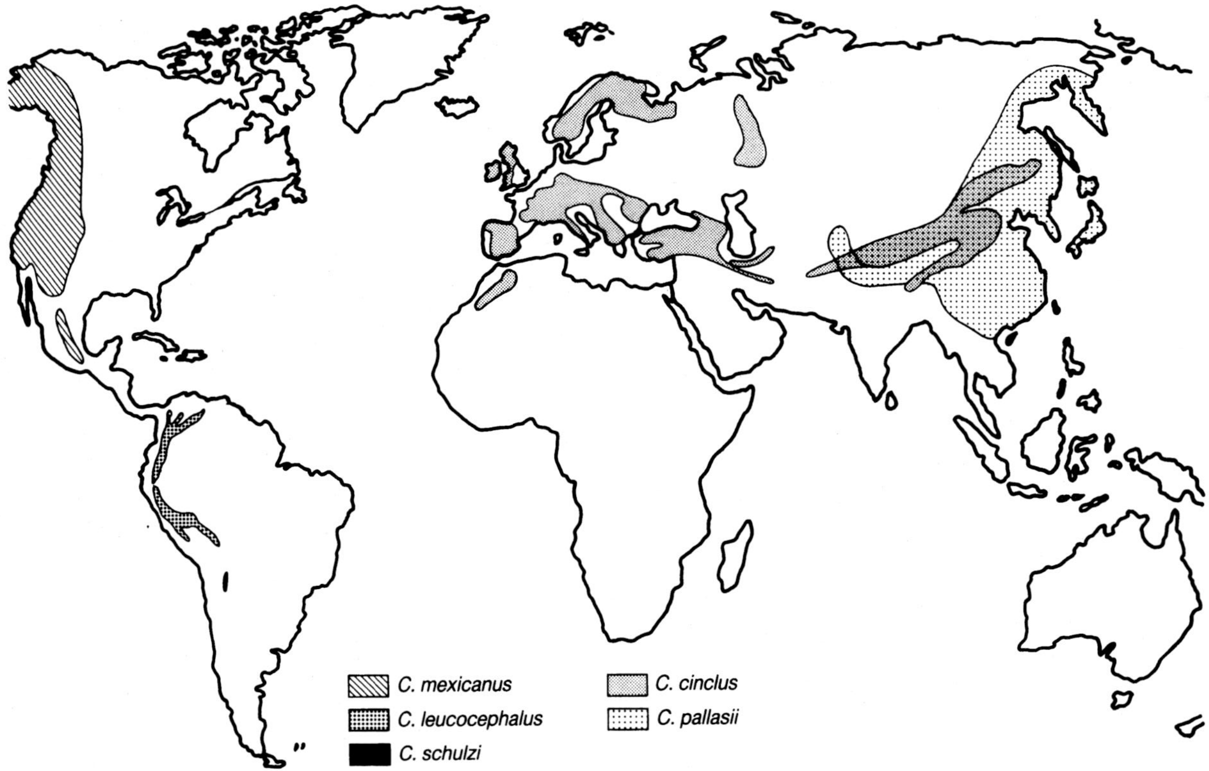

Figure 1. *Distribution of the five species of dipper.*

Brown Dipper C. pallasii *river in Nepal*

(Turdidae) were the dippers' ancestors. In the currently favoured taxonomic list, the so-called Voous order, dippers are placed between wrens and accentors (Prunellidae).

Steineger (1885) wrote that 'in appearance, movements and nest-building, etc., it (sic. the dipper) is a gigantic wren' and that 'South America is the cradle of the genus', with the neotropical forms being most like the ancestral stock. He made these assumptions believing that the white underside of White-capped Dippers from the Andes represented the primary young plumage, but had then not seen young Asian Brown Dippers.

Twenty years later he had changed his mind and professed that thrushes were the dippers' closest allies (Steineger, 1905). The telling character for Steineger was the rather spotted juvenile plumage of Brown Dippers, a similar plumage to that found in young thrushes and wheatears. He also described the young Brown Dippers as 'startlingly like enormous, overgrown fledglings of *Sialia* (bluebirds), except that they do not have the blue color on tail and wing feathers'. He believed the bluebirds of North America to have close affinities with the Himalayan Bluebird, both belonging to the Old World nightingales, bluethroats and such birds, i.e. present-day Turdidae. Steineger thus concluded that North American bluebirds and the genus *Cinclus* were of Palearctic origin. He suggested that the ancestral stock of a uniform coloured dipper *Cinclus unicolor* arose in the mountains of China during the early part of the Tertiary period. From Asia they spread across the land bridge in the Bering Sea which connected eastern Asia with North America. The massive uplift of rocks which formed the long mountain chain from Alaska down to Patagonia provided plenty of suitable mountain rivers for dippers to move south, speciating into *C. mexicanus*, *C. leucocephalus* and *C. schulzi*. Meanwhile back in Asia, the black and white or pied form evolved, and as the land rose to the west, dippers were able to expand their range and enter south-east Europe by way of Asia Minor. Steineger went on to suggest that some went south-west reaching the Atlas Mountains of Morocco, and across a land bridge to Sardinia, Corsica and into the Pyrenees, whilst others moved north-west into central Europe and thence to Britain and Scandinavia.

More recently, Sibley (1970, 1972) and Sibley *et al.* (1974) have confirmed from an analysis of egg-white protein from dipper eggs that the genus *Cinclus* did arise from some turdine (thrush) stock. Sibley, however, thought that North America was the most probable country of origin rather than eastern Eurasia. From America, dispersal may have taken place west to Western Eurasia and east and south to South America. Further confirmation of the close relationship between dippers and thrushes comes from DNA fingerprinting, which places the family Cinclidae close to Turdidae and to Sturnidae (starlings) (Cramp, 1988). The proposed classification of the Passeriformes by Sibley and Ahlquist (see Campbell and Lack, 1985) places the Cinclidae in a superfamily with the family Sturnidae (starlings) between the Bombycillidae (waxwings) and Turdidae.

White-capped Dipper Cinclus leucocephalus *in Ecuador*

The exact area of origin and the way in which dippers have speciated may never be certain, but it is interesting that dippers never spread into subSaharan Africa, nor across the United States to apparently suitable streams in the west. No dippers occur in either Australia or New Zealand. Possibly the large expanses of unsuitable habitat — desert, dry lowlands or sea — deterred dippers from colonising these areas, although some northern races of the Dipper are capable of long flights of over 1000 km. On the mountain streams in all these 'dipperless' areas other species of birds, or other fauna, occupy the dipper's niche, feeding on aquatic invertebrates. For example, the Half-collared Kingfisher frequents forested mountain streams in Ethiopia, whilst mountain torrents in New Zealand are frequented by the Blue Duck and those in north America and Iceland by the splendid Harlequin Duck.

Subspecies or Clines

Although there is now general agreement that there are five species of dippers, there is considerable confusion over the number of subspecies or indeed the existence of different subspecies, particularly of the White-throated Dipper.

A subspecies, defined as a population of which members can be morphologically distinguished, if sometimes only on average, from members of other populations of the species to which all belong (Campbell and Lack, 1985), is synonymous with a race or geographical race. Many described subspecies are, however, only samples from a population continuum showing clinal variation, i.e. a geographical gradient in one or more characters. The validity of such subspecies is therefore highly questionable, although it is better justified where forms are separated by geographical barriers.

The Cinclidae includes species, as the two South American dippers, with just one (*schulzi*) or a small number (*leucocephalus*) of clearly defined geographical races, and those with a very wide range of forms which are much less clearly defined geographically (*cinclus*). Thus the White-capped Dipper has three geographically isolated and morphologically different populations (Table 1), which can justifiably be given subspecific names. The races or forms of the (White-throated) Dipper are much more problematical. This species really shows clinal variation in size and in plumage characters, particularly in the extent and intensity of the chestnut or rufous in the band between the white belly and dark throat and upper breast. There are slight variations too between populations in the colour of the head and nape and in the width of the dark feather fringes on the underparts, although bleaching and wear of plumage can of course also greatly affect both colour and feather fringes.

Despite the clinal gradient in the Dipper, taxonomists have in the past recognised around two dozen races. These are not always geographically isolated; nor are they all clearly defined morphologically, there sometimes being considerable variation within any one population. Thus whilst the striking white-bellied race (*leucogaster*) of the Dipper from Eastern Siberia and Turkestan is very distinctive with white from the throat to the lower belly and whilst Dippers of the large nominate dark-breasted form in Scandinavia (*cinclus*) are fairly uniform in appearance, others in parts of central Europe (*aquaticus*) can be extremely variable with dark-breasted and rufous-breasted forms occurring together in the same area. Balat (1961) has described a population of Dippers in Czechoslovakia where those birds living in steep, damp, well-vegetated valleys are dark brown, and those in adjacent but wider, more open and warmer valleys are rufous-brown. Cramp (1988) suggested that these colour differences may apply to the whole of the Western Palearctic with dark brown birds inhabiting areas with cooler, wetter climates and rufous-brown birds in warmer, drier areas. Within a population too, males may be much darker than females and older birds darker than young (Richter, 1954a; Balat, 1961). There may also be marked size

Rufous-throated Dipper Cinclus schulzi *in the hand*

differences between young and old birds, between the sexes, with relative sexual differences increasing towards the north (Schmid and Spitznagel, 1985), and between individuals (Ormerod, Tyler and Lewis, 1986).

Not only may the same climatic conditions produce similarly coloured birds but also similarly sized birds. Generally northern populations such as the nominate *cinclus* in Norway and Finland are larger than races further south. This has been explained by Bergmann's rule (Bergmann, 1947), birds being larger in the cooler north. Schmid and Spitznagel (1985) found in southern Germany that there was a geographical variation in body size in four different populations, which they believed conformed with Bergmann's rule. Some races further south do not conform to any latitudinal pattern. Thus *persicus* in Iran and *minor* in Morocco are unusually large, but this can be explained perhaps by the low temperatures at the high altitudes at which Dippers occur there (see James, 1970). Dippers that we weighed and measured in the Atlas Mountains of Morocco proved to be larger and heavier than even the northernmost *cinclus* (Tyler and Ormerod, 1991). The subspecific name *minor* Tristram 1870 was given to the North African Dippers on the basis of a small female collected in Algeria.

Table 1. Races of the five species of dipper

Race	*Range*	*Plumage/size characters*
EURASIAN or WHITE-THROATED DIPPER *Cinclus cinclus*: Western Europe and NW Africa, across Asia to W China (after Cramp, 1988)		
hibernicus (Hartert 1910)	Ireland, Outer Hebrides, W Scotland	Darkest of all West Palearctic races; narrow rufous band between breast and belly
gularis (Latham 1801)	Scotland (excl. W), Wales, N, W and C England	Dark brown above dark belly with pronounced and broad chestnut-brown band on lower breast and central belly
cinclus (Linnaeus 1758) (incl. *pyrenaicus*, *sapsworthi* and *atroventer*)	Scandinavia, W and C France, NW Spain, Corsica, Sardinia	'Black-bellied', i.e. no rufous band on lower breast and belly
aquaticus (Bechstein 1803) (incl. *meridionalis*, *orientalis*)	C, S and W Europe: Belgium, E France, S and E Spain, Italy, Sicily, Poland to Greece	Pale above, grey-brown head and nape; bright rufous-chestnut on lower breast and belly
minor (Tristram 1870)	Morocco, Tunisia, Algeria	Similar to *aquaticus* in colour; larger
caucasicus (Madar/asz 1903) (incl. *amphitryon*)	Caucasus, Iran, Iraq, Turkey	Dull grey-brown head, nape, mantle, breast and belly; grey rump and upper tail coverts; some have tinge of rufous on breast/belly
persicus (Witherby 1906) (except for *minor*)	SW Iran in Zagros Mountains	Larger and paler than any W Palearctic race
rufiventris (Tristram 1884)	W Syria, Lebanon	Similar to *persicus* but smaller, with a darker rufous breast and belly band
olympicus	Formerly in Cyprus; extinct	Similar to *caucasicus*
uralensis (Serebrovski 1927)	Eastern Europe; Ural Mountains, much of the former Soviet Union	Intermediate in colour of upper parts between *cinclus* and *aquaticus*
leucogaster (Bonaparte 1850)	N Afghanistan, NE through Turkestan and SE Siberia to Sea of Okhotsk	'White-bellied Dipper'. The only race with a white belly
cashmierensis (Gould 1860)	Throughout Himalayas, rare Nepal	Chocolate-brown upper parts and lower belly; pale shading on upper belly
przewalski (Bianchi 1905)	Mountains of W China, SE Tibet and N Bhutan	Easternmost race similar to *cashmierensis* but with a wholly dark belly

BROWN DIPPER *Cinclus pallasii*: SW Siberia and Turkestan across the Himalayas to Japan		
tenuirostris (Bonaparte 1850)	C Asia	Chocolate-brown
dorjei (Kinnear 1937)	S Assam, N Burma, N Thailand, SE Tibet	Dark chocolate-brown; young washed with rufous
marila	Khasia Hills, India	Included with *dorjei* by Ali and Ripley (1983)
pallasii (incl. *hondoensis*: Japan)	NE Asia, Japan, W China, N Thailand, N Vietnam	Dark chocolate-brown
AMERICAN DIPPER *Cinclus mexicanus*: Aleutian Islands and Alaska, Canadian Rockies, West USA down to Costa Rica and Panama		
unicolor	W Canada, W USA	Grey-brown head and paler plumage than the four races below
mexicanus	N and C Mexico	Varying in darkness of plumage
anthonyi	S Mexico and Guatemala	
dickermani	S Mexico	
ardesiacus	Costa Rica and Panama	
WHITE-CAPPED DIPPER *Cinclus leucocephalus*: Sta Marta Mountains of Colombia down Andes to Peru and NW Bolivia		
leucocephalus	Peru and NW Bolivia	Black/blackish with white crown, throat and upper breast; white wing bar
rivularis	Sta Marta Mountains, Colombia	White of throat spotted with grey; as *leucocephalus* but paler
leuconotus	S Colombia, W Venezuela, Ecuador	Small size; white patch on back and white underparts except lower flanks and vent
RUFOUS-THROATED DIPPER *Cinclus schulzii*: SW Bolivia and mountains of Tucaman, Catamarca and Jujuy in Argentina		
schulzi		Dark grey; light grey chin; rufous throat; white patch on inner webs of primaries more extensive than in *leucocephalus*

The largest race of the Dipper C. c. minor*! in Morocco*

Clearly, then, there is a danger in determining races particularly from a very small number of museum specimens. These in any case may have faded with age and certainly shrunk. It would be very useful to have data from live birds from those areas such as Lebanon and parts of Iran where data are particularly scant. Sadly though political considerations may not permit this for some time!

For convenience, it is undoubtedly useful to ascribe subspecific names to populations. Formerly numerous races were recognised but Greenway and Vaurie (1958) lumped a number of similar adjacent races together. Cramp (1988) further modified Greenway and Vaurie's list through more amalgamations, accepting some 12–14 races (Table 1). For further descriptions and discussions of the so-called races of the Dipper, the reader is referred to Greenway and Vaurie (1958) and Cramp (1988).

Table 2. Population estimates and status of the Eurasian or White-throated Dipper

Country	*Estimated population*	*Status*
Britain and Ireland	30,000 (Sharrock, 1976) Up to 26,000 (Gibbons *et al.*, 1993)	No marked change but some local declines in Wales and Scotland due to acidification (Ormerod and Tyler, 1987b; Vickery, 1991); other increases as in Hebrides (Tyler and Ormerod, 1993). Bred pre-1950 in Isle of Man and Orkneys; now extinct
Norway	10,000–15,000 (Efteland, pers. comm.; see Norwegian Atlas)	Concentrated in south, but under-recorded in north
Sweden	10,000 (Ulfstrand and Hogstedt, 1976)	Some declines due to acidification (Eriksson, pers. comm.)
Finland	400 (Merikallio, 1958); 300 (Koskimies, 1989)	Rare; northern range has markedly diminished due to damming and logging (Koskimies, 1989)
Denmark	Six breeding records pre-1950; none bred 1951–59; 6–8 (Cramp, 1988)	Common winter visitor (Preuss, 1959) but Roos (1982) described it as only wintering in small numbers
Belgium	1200 (Lippens and Wille, 1972); 740 (Jacob, 1988)	Winter visitor too (Evers, 1982)
Holland	Only bred 1910–15; 1933	Winter visitor only (Evers, 1982)
France	1000–10,000 (Yeatman, 1976; Alsace 100–200 (Kempf, 1977)	No longer breeds in Normandy or Bretagne
Austria	Unknown but common	Local losses due to power stations and loss of old bridges (Wagner, 1985)
Switzerland	Unknown	Not uncommon; no obvious change
Czechoslovakia	Unknown; not uncommon	No change reported
West Germany	5700–6600 (Rheinwald, 1982); 7300–32,200 (Cramp, 1988)	No major change but some declines due to acidification (Kaiser, 1988)
East Germany	Unknown	Scarce and declining (Cramp, 1988)
Poland	Unknown	Now scarce; declining in some areas (Cramp, 1988; Wiatr, 1969)
Hungary	10–15 (Horvath, 1988; Wiatr, 1969)	Rare and declining

Waterfall nest site in Mid Wales uplands

Former USSR	Unknown	Common in west and Caucasus (Dementiev and Gladkov, 1954); breeds occasionally in Estonia and Latvia where mainly a winter visitor. No information for eastern populations but not uncommon
Yugoslavia	Unknown	
Bulgaria	Unknown	Common in upland areas
Iran and Iraq	Unknown. Vagrant to Iraq (Keith, Urban and Fry, 1992)	Frequent on streams near Teheran and Alborz (Meiklejohn, 1948)
Turkey	Unknown but estimate by ICBP/DHKD* of 500–5000 (unpubl.)	Sparsely distributed (Vittery, Porter and Squires, 1971)
Greece	Less than 50 (Cramp, 1988)	Scarce
Italy and Sicily	Unknown	Decline in Sicily due to water regulation, notably channelisation (Sara and Sorci, pers. comm.)
Spain, Corsica and Sardinia	Unknown	Possibly local declines due to water regulation
Malta and Cyprus	None (Flint and Stewart, 1983)	Extinct on Cyprus; occasional visitor to the islands
Portugal	Unknown	Possible declines in some areas; due to water regulation and deforestation?
Morocco	Min. 200, probably 500–1000 (Tyler and Ormerod, 1991)	Widespread in Middle and High Atlas
Algeria	Less than 100? Rare	Only known from 5 localities (Keith, Urban and Fry, 1992)
Tunisia	None?	Vagrant (Keith, Urban and Fry, 1992)
Libya	None?	1 claim 1916 (Keith, Urban and Fry, 1992)
Afghanistan		Apparently common
Pakistan, India, Nepal	Unknown	Frequent in the Himalayas but scarce in Nepal (Ali and Ripley, 1983); concern over possible impact caused by deforestation and silting of rivers in some areas (Tyler, 1994)

NOTES:
*ICBP = International Council for Bird Preservation (now BirdLife International). DHKD = Turkish Wildlife Conservation Society, Doğal Hayati Koruma Derneği.

STATUS OF THE FIVE SPECIES OF DIPPER

Threats to dippers are described in Chapter 10. Suffice it to say here that some island populations are already extinct (Cyprus: Flint and Stewart, 1983) whilst others are at risk or show alarming declines due to various river works such as hydro-schemes or channelisation (Sicily: Sari and Sorci, pers. comm.), as well as from drying up of rivers elsewhere in the Mediterranean region due to a combination of deforestation, drought and profligate use of water for agriculture and tourism. Mainland populations of all species are also locally adversely affected by river works, by reservoir construction or overabstraction for irrigation, by pollution in various forms and by general habitat degradation. Some current estimates of the populations of the Dipper in various countries are shown in Table 2.

Dipper swimming and second Dipper on rock with water droplets on back

CHAPTER 2

Riverside life: morphological and physiological adaptations to a special environment

Rivers are special environments. They provide abundant food, but pose many constraints such as the force and turbulence of rushing water — a medium whose refractive index differs from air — low temperatures and slippery rock surfaces. Dippers clearly need special attributes to survive in such a hostile environment.

MORPHOLOGICAL ADAPTATIONS

Dippers are unique amongst passerine birds in their ability to dive and swim underwater, and to exploit aquatic invertebrates as prey. At first glance their general appearance gives little clue as to how they can lead an aquatic or semi-aquatic existence. Although they have no obvious adaptations such as webbed feet, they are however superbly adapted to their lifestyle. They are robust and perfectly streamlined, with a number of morphological and physiological adaptations which enable them to live in and by cold mountain rivers. Compared with non-aquatic passerines, their plumage is very dense (Goodge 1960). This helps to insulate them against the low temperatures of high altitudes and often ice-cold water, and also enhances evaporative cooling in hot weather. When temperatures are high, dippers may cool themselves by standing in water to increase the loss of heat through the uninsulated legs and feet (Murrish, 1970a), although in north-west Argentina Rufous-throated Dippers, along with *Cinclodes* species, avoid high temperatures around midday by seeking shade below rock overhangs by the river. On dull or wet days they are more evident on rocks out in the river.

Dipper feathers are heavily waterproofed, so that when they emerge from a river, water rolls easily off their plumage in droplets and does not wet the feathers and cause chilling. This waterproofing is achieved by frequent preening which coats the feathers with oil from the unusually large preen gland situated at the base of the tail. The stout, strong bill is ideal for probing among rocks and winkling out prey. The feet are not webbed but are large with strong toes and claws which enable the bird to grip rocks and pebbles on the river bed. Their legs are long and strong too, and are used to swim strongly in a paddling motion. The wings are relatively short for a bird of its size, but have a very well-developed

A Dipper C. c. gularis *with water droplets rolling off plumage (Rick Packwood)*

musculature. This enables the bird to use its wings as flippers when underwater, beating them to propel it down to the river bed, and maintain its position under the water. Dippers have nasal flaps like other aquatic birds to prevent water entering their nostrils (Grinnell and Storer, 1924).

PHYSIOLOGICAL ADAPTATIONS

Murrish (1970b) showed that dippers also have many of the physiological adaptations for diving seen in other aquatic birds. Upon submersion they demonstrate an immediate drop in heart rate followed by a further gradual decline. On surfacing, the heart rate rapidly increases again. Dipper blood has a high haemoglobin concentration and hence a greater capacity to store oxygen compared with blood of terrestrial birds.

Goodge (1960) was interested in how dippers were able to see prey equally well both above and below the water. The cornea of the eye has an important refractive role in air but does not usually work underwater because its refractive index is similar to that of water. Goodge studied refraction in the dipper eye using a retinoscope and by stimulating accommodation of the lens electrically or chemically. He also made a histological study of eyes of American Dippers to try to identify any structural modifications which might account for amphibious vision in dippers, comparing these with the eyes of related species such as the American Robin, Varied Thrush and House Wren. The results of the retinoscopy confirmed that dipper eyes had enhanced powers of accommodation. The only part of the eye that differed significantly from those of the other species examined was the iris, which had a very well developed sphincter muscle. Goodge suggested that pressure exerted by this muscle could change the curvature of the lens, so enabling the greater powers of accommodation. A similar well-developed iridial sphincter had been found in another group of diving birds, the cormorants, and in marine turtles by Hess (1912).

BEHAVIOUR PATTERNS OF DIPPERS

Two behaviour patterns are characteristic of dippers. Firstly, dippers are unusual in their dipping behaviour, which is shown by three of the five species, and secondly, they frequently blink. They make the characteristic dipping or bobbing movements of the body in a variety of situations. They dip when resting on a rock or between bouts of feeding, although dipping becomes more vigorous during territorial encounters, during courtship and when the bird is alarmed. The two South American species repeatedly flick their wings instead of dipping, at the same time depressing their tails. Other riparian species of bird such as

Common Sandpipers share the bobbing habit with dippers, although in sandpipers the whole body is not dipped and movements are not as frequent as in dippers. Most other species of birds that live along rushing rivers *either* move their tails from side to side or up and down (e.g. wagtails, forktails, shaketails, Plumbeous and White-capped Water-redstarts, Torrent-larks and Riverine Flycatchers), *or* flick their wings. It has been suggested that the function of dipping, wing-flicking or tail-wagging is to render the bird less conspicuous to predators against a background of moving, often turbulent, water. These movements however, probably act as intraspecific advertising or contact signals adapted to overcome the difficulty of communication against the background noise of their environment (Orenstein, 1975). In some situations dipping in an exaggerated way is clearly a means of signalling to other dippers, with ritualised dipping a feature of both threat and courtship displays.

The calls made by all five species of dipper are remarkably similar, being loud and high-pitched; the calls are also similar to those made by many other birds that live on torrential, noisy rivers. The vocalisations of these river birds are concentrated within a narrow frequency belt (4.0–6.5 kHz) above that of the torrent noise (<2 kHz) so that long-distance communication is possible (Martens and Geduldig, 1990).

Why do dippers in Britain, Eurasia and in South America have a striking dark brown and white plumage, and yet have a more uniform colour elsewhere? The white bib of the (White-throated) Dipper and white cap and belly of the White-capped Dipper are conspicuous at close quarters. However, the white patches on these two species of dipper might help break up the outline of the birds and actually render them less conspicuous to predators, especially on torrential mountain rivers, with much 'white' water. Only in some situations as on lowland rivers does the white breast stand out. In some circumstances as during aggressive and courtship behaviour the white plumage is used as a signal; it may also serve with dipping and blinking to signal alarm, to warn other dippers of a predator's approach or as a predator signalling strategy. Some other stream-adapted species such as Torrent-larks and forktails also have a striking and contrasting plumage, which Orenstein (1975) argued was, along with body, wing or tail movements and penetrating calls, to aid intraspecific communication. The drab colour of Brown, American and Rufous-throated Dippers does not appear to conform to this hypothesis, although the white wing flash of the South American birds may well be a means of signalling to other dippers.

The white underside of the White-capped and (White-throated) Dipper may serve to render them less conspicuous to their aquatic prey, the same pattern being shown by diving auks. If this is so, it is perhaps all the more strange that the Brown and American Dippers are uniformly dark in colour, yet are clearly successful predators. When underwater, though, the plumage of dippers appears silvery grey due to a mass of small air bubbles trapped in the feathers; this may make them less conspicuous to mobile prey on the river bed.

The characteristic and unusual frequent blinking of the eyes appears to be

Wing of Rufous-throated Dipper showing white bar

special to dippers. The eye-blinking was for some time wrongly attributed to movements of the nictitating membrane or third eyelid by many observers such as Grinnell and Storer (1924), Cordier (1927), Rankin and Rankin (1940) and Hann (1950). It was not until James Alder (1957) observed birds that he had tamed, that it was recognised that there were two distinct types of eye movement: the quick blinking of the whitish upper eyelid, which may or may not accompany dipping, and which is the conspicuous movement seen by observers; and a high speed flicking of the nictitating membrane, which is only evident at close quarters. Goodge (1960) then observed a number of captive American Dippers and showed too that it is the movements of the white-feathered lids, mainly the upper lid, which produces the white flash seen in the field. When a dipper is at rest, the small white feathers covering the upper lid are concealed by the larger dark head feathers, but the white marginal feathers of the lower lid may be visible. Other species of birds lack feathers covering the lids although they do have rows of small dark feathers on the margins of each lid.

Goodge (1960) found that American Dippers could blink either both or just one eye, and that sometimes they might blink as many as 40–50 times per minute. In Irish Dippers, Rankin and Rankin (1940) recorded similar blinking rates, 36–54 blinks per minute, which could increase when birds were dipping vigorously. They noted that blinking always accompanied dipping but not vice versa. The white flashing of the eyelid is believed to be a signal (Creutz, 1966)

Structure of eye showing nictitating membrane and white eyelids

used in courtship and threat displays. At these times dippers blink and dip more frequently, but they also do so when alarmed as when a human observer approaches too closely.

The nictitating membrane in dippers has been described as milky white and cloudy (Goodge, 1960) or as bluish-white and semi-transparent (Alder, 1957). Other birds such as owls and certain passerines also possess a cloudy rather than a clear nictitating membrane but the reason for the cloudiness is unknown. Dippers may move the membrane across the eye at the same time as they blink the eyelids, or independently of this. The membrane moves from the anterior dorsal surface in a posterior ventral direction across the eye (Goodge, 1960). Movements of the nictitating membrane across the surface of the cornea presumably help keep the eye free from particles of dirt, or remove water droplets when the bird is on land; it may also serve some protective function when the bird is underwater.

These various adaptations and strange behaviour patterns make the family Cinclidae unique amongst passerines. Dippers are land birds but are able to swim and survive in the relatively hostile environment of mountain streams and rivers. Their combination of attributes has attracted many people, and led some, not least John Muir in 'The Mountains of California', to immortalize dippers in prose and poetry.

UNDERWATER BEHAVIOUR

The unusual aquatic lifestyle of dippers puzzled observers for some time. In particular, how dippers were able to move underwater was a subject of long debate earlier this century. Much was written about both the European and

North American species, but there was considerable controversy about whether dippers were able to walk underwater along the bottom of a river without using their wings. Coward (1969) did not believe there was any truth in the assertions that a dipper could defy the laws of specific gravity and walk on the river bed in this way. However, Madon (1934) and Ingram, Salmon and Tucker (1938) insisted that although dippers may occasionally use their wings to maintain balance or in strong currents, they normally walked about, sometimes completely submerged, on the river bed, with no wing movement. They quoted their own observations of dippers walking in this way in any direction in relation to the current, turning over pebbles and only rarely flicking their wings. Ingram, Salmon and Tucker were unsure as to whether dippers gripped on to the stones on the river bed with their feet. Although they did not observe dippers swimming underwater using their wings, they quoted Arthur Brook as having seen birds swim in this way in deeper water.

De Vogue (1934) suggested that a dipper was able to maintain its position on the river bed by lowering its head and allowing the force of the current to play against its back. An experiment with a wooden dipper was carried out by Dewar (1938) to support this hypothesis, but Ingram (1938) doubted the results. Brownlow (1949) also discounted this experimental demonstration. He described his own field observations in which birds walked across the current, or obliquely to it, for periods of up to 18 seconds and for distances of up to 2 metres, and concluded that dippers normally rely on footholds to keep them down. He believed they only used their wings to swim down to the bed of the river or up to the surface if the water was deeper than 'a few inches', to keep them down and regain footholds if they lost their grip, to search for food if the bottom had no suitable footholds or when there was an urgent need to return quickly to the bank.

These early observations were hampered by the authors' not always being able to see clearly what the dippers were doing underwater. In 1952 Jones and King were fortunate to be able to watch a Dipper which visited their salmon observation tank on a tributary of the River Dee in Wales. They were also able to take photographs to record the bird's activities. The Dipper frequently dipped its head underwater, presumably to see if there were any salmon eggs visible around the rocks on the floor of the tank (Jones and King had baited a rock with eggs for the bird). It descended into the water, either by walking or diving, and then swam actively about the tank, even swimming upstream against the current, by means of rapid wing movements. It was able to stay submerged in the 20–30 cm of water for up to 30 seconds. Jones and King did not observe it grip the stones in the tank using its feet, but thought it may push itself off the bottom with its feet. They did, however, observe the bird's silvery appearance when underwater, due to a film of small air bubbles trapped on the surface of the plumage. This had also been controversial: some earlier authors claimed to see such a film, others such as Ingram *et al.* did not. Cordier (1927), writing of the American Dipper, disputed that birds ever showed a silvery appearance underwater.

Underwater shot of a Dipper (Oxford Scientific Films)

Brownlow (1953), after reading of Jones and King's clear observations of a Dipper using its wings underwater, accepted that his earlier observations might not have been entirely correct. He accordingly modified his views, saying that dippers walked when in shallow water and swam in deeper water. Bagnall-Oakley (1954) contributed further observations of a dipper using its wings, particularly when diving down, its head held forwards and downwards and its bill lowered, or when making a spurt to catch prey. He noted that the bird moved in any direction and emerged by stopping and bobbing up 'with a cork-like buoyancy'.

Goodge (1959) who carried out laboratory experiments using the American Dipper, proved conclusively that wing movements were essential for the dipper to maintain its position underwater and also for propulsion, and that the role of the legs was secondary to that of wings underwater. The legs were used, in a running motion, to assist the bird's progress on the bottom or to briefly cling to the bottom, as well as for paddling or swimming on the surface, but there was no evidence from Goodge's tank experiments that American Dippers could walk underwater without the aid of their wings and he discounted earlier observations of their being able to do so.

An observation by Broadley (1985), who caught a one-legged Dipper of normal weight on two occasions in January at an interval of over three weeks, showed that the legs must be much less important than wings for feeding in water. More recent films of dippers feeding underwater, such as those made by Oxford Scientific Films, have clearly shown their wing movements, and also that birds sometimes grip to the substrate with their strong feet and claws to help prevent their floating back up to the surface.

When diving, dippers do not remain underwater for long. Goodge (1959) thought that American Dippers probably did not dive for longer than 10 seconds, although Muir (1894) implied that they could remain underwater for as long as two or three minutes. Bagnall-Oakley (1954) recorded dives by the European Dipper lasting 17–23 seconds. In SW Norway, the longest dive timed by Efteland (1976) was 14 seconds, whilst Cordier (1927), Jones and King (1952) and Vader (1971) noted dives of almost 30 seconds. Longer dives of up to a minute are claimed by Graf (1968) and Wesenburg-Lund (in Curry-Lindahl, 1963). Such long dives appear unlikely. The more usual average length of dives seems to be in the range of 4–20 seconds (Brehm, 1907; Curry-Lindahl, 1963; Creutz, 1966; Hafthorn, 1971; Vader 1971) but Efteland (1976, 1983c) found the average duration to be only 3.2–3.6 seconds.

Dippers usually make a series of dives before resting or switching to another form of foraging. Efteland (1983c) found that such a series of dives lasted for 30–1170 seconds, with a mean of 297 seconds, and within a series there were 2.8–11.5 dives per minute, with a mean of 5.1. Efteland also found that the duration of dives increased with depth of water, as too did their frequency.

Not all species of dipper, however, swim and dive. The two South American species rarely, if ever, dive underwater although they often wade in rivers up to their bellies or stand on the lips of cascades and waterfalls, picking prey from rocks and often dipping their heads below the surface and becoming temporarily submerged in rushing water. The explanation for the differences in behaviour between the White-capped and Rufous-throated Dippers and the other species of dipper is not known but possibly competition with Torrent Ducks in the Andes may have led the South American species to forego an underwater lifestyle. Their foraging behaviour is rather similar to that of Little Forktails which share Himalayan rivers with Brown Dippers. The Little Forktails do not swim or dive, whereas the large Brown Dippers frequently do so, plunging even into raging torrents.

Dipper in riffle with caddis larva in its bill

CHAPTER 3

Factors influencing the distribution and breeding abundance of dippers

There can be few better places to be than by an upland stream with mossy, spray-drenched rocks, waterfalls and overhanging alders or oaks. What determines your chances on any particular stream of seeing a Dipper flash by or perch on a mid-stream boulder, then dip and plunge into a pool? What makes one river a good Dipper river and another, apparently similar, poor for Dippers? In this chapter we look at the various factors that influence the presence of these birds and their breeding abundance.

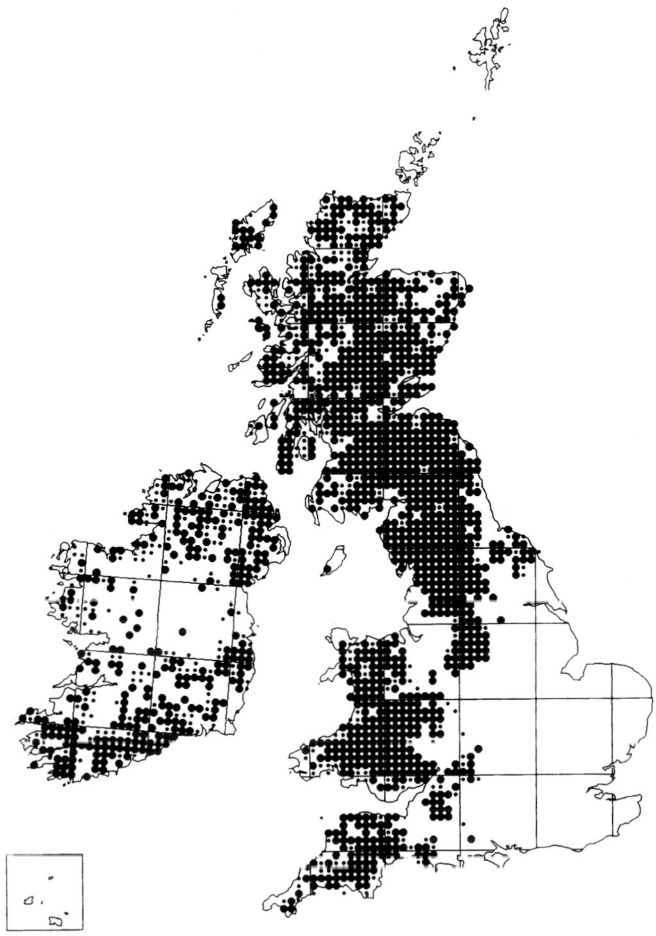

Figure 2. Distribution of Dippers in Britain and Ireland 1988–91.
From *The New Atlas of Breeding Birds* (Gibbons, Reid and Chapman, 1993) with permission from the British Trust for Ornithology. Large dots show breeding records, small dots show birds seen but no evidence of breeding. Note: one breeding record has been moved by up to two 10-km squares.

DISTRIBUTION OF DIPPERS *Cinclus cinclus*

In Britain and Ireland Dippers are closely associated with rivers and streams in upland areas, occurring throughout much of Scotland, Ireland, Wales, northern and south-west England (Figs 2 and 3). Their occurrence as a breeding bird elsewhere is limited to rivers in the Welsh borders (Shropshire, Hereford and

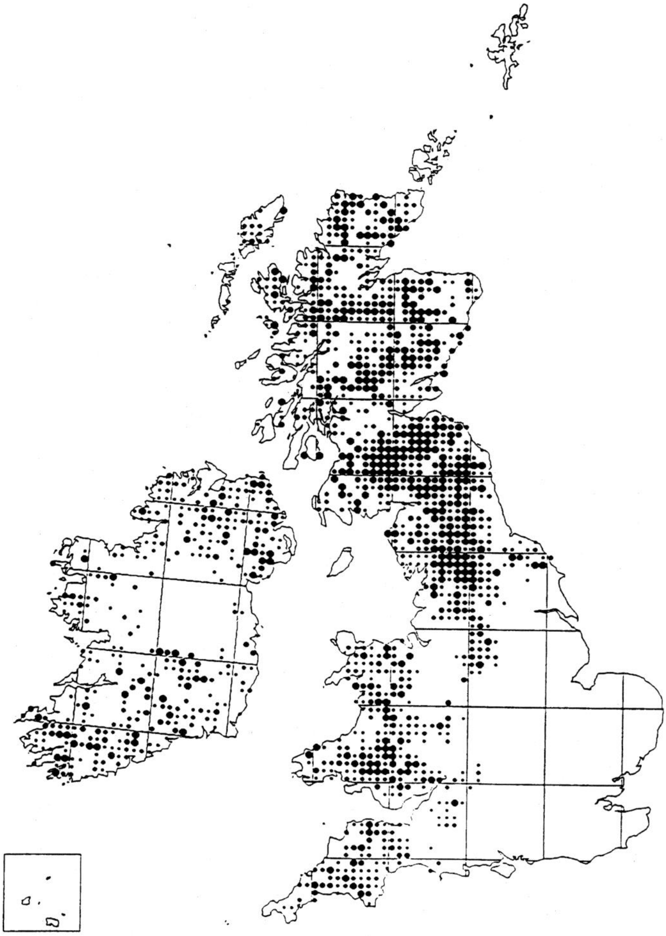

Figure 3. Abundance of breeding Dippers in Britain and Ireland 1988–91.
From *The New Atlas of Breeding Birds* (Gibbons, Reid and Chapman, 1993) with permission from the British Trust for Ornithology. Largest dots show highest numbers of Dippers. For abundance scores, see Gibbons *et al.* (1993). Note: one breeding record has been moved by up to two 10-km squares.

Worcester, Gloucestershire, particularly the Forest of Dean area), the Cotswolds (Gloucestershire and northern Wiltshire, e.g. By-Brook), Avon and occasionally to chalk rivers of southern Wiltshire and down into Hampshire. Dippers will sometimes feed by oligotrophic lake shores especially outside the breeding season, and particularly in the winter months they may forage along the coast. Migrants in Britain from Scandinavia or possibly the Continent may also winter by ponds and lakes and on slower-flowing rivers than would be the case in the breeding season. Hence black-bellied Dippers of the nominate race may occur by dykes in the fenland of East Anglia or on streams in Sussex.

Dippers occur throughout Europe and Asia and in North Africa wherever there are suitable upland rivers. They occur in boreal, temperate, Mediterranean and steppe climatic zones (Voous, 1960). Their population size in any country is related to the area of mountainous or other upland habitats and the lengths of suitable fast-flowing rivers in that country. Thus in Spain and Portugal Dippers are restricted to mountains in the north and in the south, and absent from the central plains, whereas in Switzerland they occur throughout, breeding from 200 m to over 2000 m on rivers flowing through forests and through towns and cities (Glutz von Blotzheim, 1962). In Russia, there are isolated Dipper populations wherever mountain masses occur, with birds breeding up to 2500 m but only wintering where open rivers or ice-free stretches occur (Dementiev and Gladkov 1954). In Morocco, they occur only in the mountains of the Middle and High Atlas, where they breed and winter on permanent rivers. Temporary rivers may dry before breeding is completed, and even in the winter the temporary rivers have unstable beds and contain few invertebrates by contrast with permanent watercourses. These temporary rivers may, however, be important during dispersal and at some other times of the year.

A good Dipper Cinclus c. gularis *river in Wales*

FACTORS AFFECTING BREEDING ABUNDANCE

Pairs of Dippers establish linear breeding territories along a river, with the length of a pair's territory varying considerably from just 300–400 m to over 2500 m (Table 3). Territories of adjacent pairs usually abut but they may, on less favourable watercourses, be separated by a stretch of unoccupied river. Within its territory, a pair must clearly have a good nest site and safe roost sites, but above all they need a sufficient area of river from which to obtain enough food for themselves and one or more broods.

The breeding abundance and territory length of Dippers has frequently been related to factors such as altitude, gradient and the presence on the river of rocks, pools and riffles. A plentiful food supply, good water quality, suitable nest sites and/or a riparian fringe of broad-leaved trees have also been shown to be important factors. It is in fact a combination of certain physiographic features and an abundance of certain groups of invertebrate prey that is crucial if Dippers are to occur in high abundances.

ALTITUDE AND GRADIENT

Shaw (1978) in his analysis of the BTO Dipper nest record cards found most nests (87%) occurred below 300 m, with sites ranging from only 15 m and up to 600 m. Nests were most numerous though at altitudes of 50 to 250 m. We found a rather similar picture (Fig. 4) in the southern catchment of the Wye where 75% of nests were below 250 m and 90% below 300 m with a range of 3–650 m (Tyler and Ormerod, 1985). A correlation between breeding abundances and altitude was evident when we looked at Dippers on tributaries throughout the Wye catchment in mid- and south Wales (Fig. 5). Round and Moss (1984) had also found such a correlation on the Wye and Severn catchments, and Marchant and Hyde (1980), too, noted that Dipper breeding abundance increased with increasing altitude.

However, we believe altitude *per se* to be immaterial, provided the gradient of the watercourse is steep enough to provide the necessary physical characteristics (rocks, riffles, shallow clear water and pools) required by Dippers. Data from over 220 km of tributaries throughout the Wye catchment revealed breeding abundances of 2.9–8.5 pairs per 10 km (Ormerod, Boilstone and Tyler, 1985). The highest abundances were on the lower River Edw, the Monnow subcatchment and the Grwyne Fawr (Fig. 6). Breeding abundance on these tributaries, all of which had steep relief, was strongly correlated with gradient (Fig. 7). We found that over 40% of the variance in breeding abundance could be explained by gradient. This relationship was similar to that described on the Wye and Severn catchments by Round and Moss (1984) and on census plots throughout Britain by Marchant and Hyde (1980), who noted an increase in the abundance

Table 3. Abundance of breeding dippers as shown by number of pairs per 10 km or by territory length

	Geographical area	*Number of pairs per 10 km (mean)*	*Territory length (m)*
BRITAIN (*C. c. gularis*)			
Serle and Bryson (1935)	R. South and North Esk	5.2	
Robson (1956)	R. Eden	2.1–6.2	110–650 (mean 430)
Horn (1985)	R. South Esk and Leithian Water	3.0–3.3	671–2080; 511–4178 (mean 1570–1593)
Cowper (1973)	Esk System	1.4–2.2	
Sinton (1988)	R. Wharf, Yorkshire	7.1	318–583
Shaw (in Cramp, 1988)	Perthshire	Up to 9.7	
Shaw (in Cramp, 1988)	Galloway	Up to 7.7	
EUROPE (*C. c. cinclus*)			
Efteland and Kyllingstad (1984)	SW Norway	1.4–2.2	
EUROPE (*C. c. aquaticus*)			
Eggbrecht (1937)	Germany		1000–1500
Breitenmoser-Wursten and Marti (1987)	Swiss Alps	Overall 4.3–7.9 but up to 20 on some stretches	
Sokolowski (1964)	Poland		1000–3000
Oelke (1975)	Central-east	0.7–10 (4)	
Jost (1975a)	W Germany	1.1–3.7 (2.4)	
Klein and Schaak (1972)	W Germany	2.7	
Schmid (1985a)	W Germany	7.7	350–2000
Rockenbauch (1985b)	W Germany	Up to 23 (2.1)	
Kaiser (1985)	W Germany		1500–4000
Dick and Sackl (1985)	Austria	2.3	
Wagner (1985)	Austria	2.8	
Kaiser (1988)	W Germany	0.6–1.7	
NORTH AMERICA (*C. mexicanus*)			
Hann (1950)	Colorado		732–1609
Bakus (1959a, b)	Montana		(653–687)
ASIA (*C. pallasii*)			
Sunquist (1976)	Pakistan		315–562 (425)
ARGENTINA (*C. schulzi*)			
Tyler (1994) Tyler and Tyler (in press)	NW Argentina		600–1200

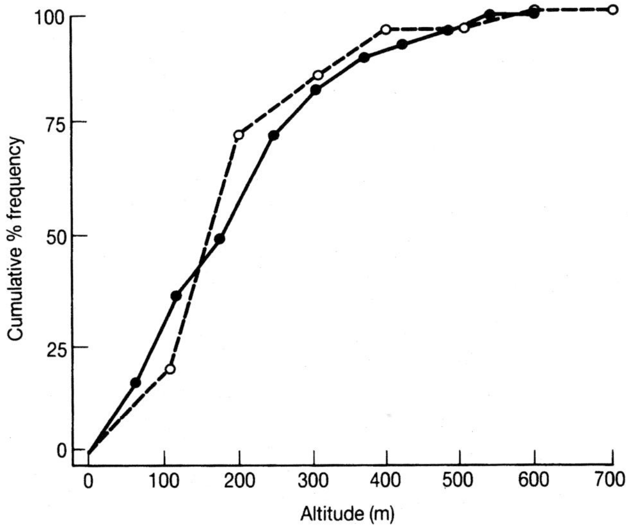

Figure 4. Cumulative percentage frequency distribution of the altitudes (metres above sea level) of Dipper nest records.
From: ○ Shaw (1978), and ● Tyler and Ormerod (1985).

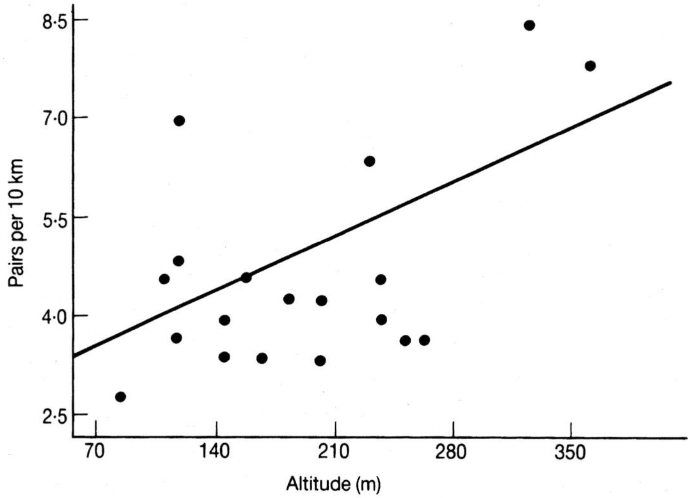

Figure 5. The relationship between the abundance of breeding Dippers (pairs per 10 km) and altitude (metres above sea level).
NOTES: The line was fitted using regression relationships between the abundance of breeding Dippers and selected physical/macroinvertebrate variables (Ormerod, Boilstone and Tyler, 1985).

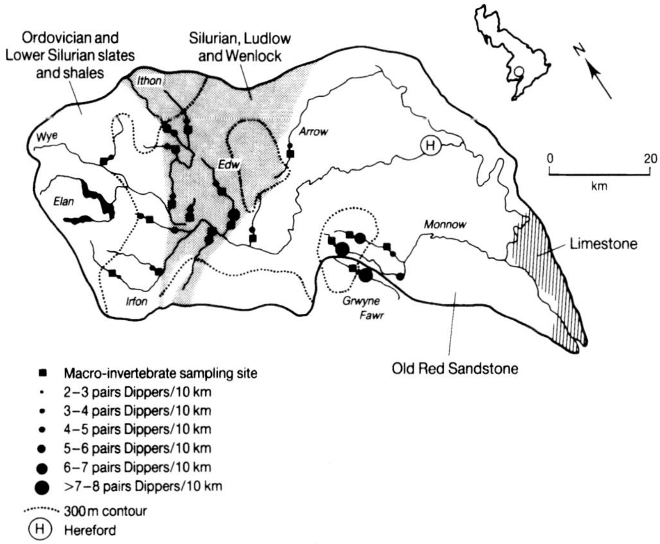

Figure 6. The Wye catchment in mid- and south-east Wales showing macroinvertebrate sampling positions and abundances of breeding Dippers recorded in 1982.

NOTES: Principal geological divisions are also shown: Ordovician and Lower Silurian (clear); Silurian, Ludlow and Wenlock (shaded); Old Red Sandstone (clear); Carboniferous (hatched) (Ormerod, Boilstone and Tyler, 1985).

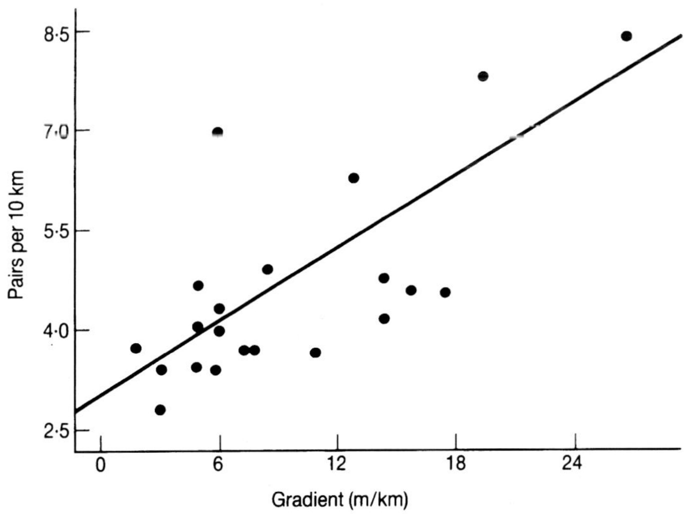

Figure 7. The relationship between the abundance of breeding Dippers (pairs per 10 km) and gradient (m/km).

The line was fitted as for Fig. 5.

of breeding Dippers from 2.5 to 8 pairs per 10 km at sites with gradients between 2.5 and 20 m/km. They found maximum abundances of 16.4 pairs per 10 km at gradients of 15 m/km or over (Marchant and Hyde, 1979).

In mainland Europe, too, many observers have shown a marked correlation between density and gradient, e.g. Dick and Sackl (1985) in Germany. Likewise in Western Spain, Peris *et al.* (1991) concluded that gradient was the most important abiotic factor affecting distribution. There, Dipper breeding abundances on streams (which were not acidic) ranged from 1.4 to 15 pairs/10 km, with the highest abundances at gradients of 105 m/km.

As it is the higher reaches of rivers that generally have the steeper gradients, Dipper abundances are inevitably usually greater at higher altitudes. The distribution of breeding pairs thus generally reflects the distribution of watercourses with steep gradients, whether these occur close to sea-level or in high mountains. That breeding birds are not restricted by altitude is shown by pairs nesting only a few metres above sea-level. For example, Dippers regularly breed at 3 m above sea-level on the Angidy Brook at Tintern in Gwent, just above its confluence with the estuarine section of the lower Wye; this brook has a steep gradient of 50–60 m/km and is consequently rocky with shallow water, riffles and pools throughout its length, and supports a high breeding abundance of Dippers throughout.

Da Prato and Langslow (1976) found that in Midlothian, Scotland, the number of Dippers per unit length of river was independent of altitude. In the Swiss Alps Breitenmoser-Wursten and Marti (1987) found the vertical distribution of Dippers paralleled the distribution of available watercourses; no particular altitudinal level was preferred although the highest territories which always lay below the timberline at under 1600 m were occupied only in years when the population was high. Surprisingly, they found that gradient had no influence on the distribution of Dippers, but this is explained by all their study streams having a steep gradient.

Water Depth and Area of Riffles

Although Dippers will dive to depths of 1–2 m, they prefer feeding in riffles, where shallow water rushes over stones and small rocks, the area of riffles on a watercourse being closely dependent on gradient. The availability of shallow water where prey is more readily accessible is clearly of importance in habitat and territory selection by Dippers.

Serle and Bryson (1935) drew attention to the Dipper's need for shallow water for feeding. Many other fieldworkers have since shown the importance of shallows and riffles. For example, Shooter (1970) showed that higher breeding abundances occurred on those Derbyshire rivers which had larger areas of riffles or shallow water irrespective of potential Dipper food. He estimated that a pair of Dippers required an area of 4000 m^2 of shallow water in a territory. Murphy

(1972) also suggested that water depth, as well as bank stability and available boulders, were the proximate factors through which a pair of Dippers assessed the necessary size of a territory, with food being the ultimate factor. Horn (1985) likewise found that availability of suitable shallow riffles was one of the two most important factors in explaining the distribution of Dippers on the River South Esk and Leithan Water near Edinburgh, the other factor being availability of nest sites. Sinton (1988) in her study area in North Yorkshire found a significant correlation between the percentage area of riffle and Dipper territory area, territories being smaller in those reaches where shallow water and riffles predominated. If territories are small, and contiguous, then the number of breeding pairs will be high.

River Width

We have found that the width of watercourses (greater than *c*. 2 m wide) of similar gradient has little effect on breeding abundance or territory length. Abundances may be as high on streams only a few metres wide as on rivers of 15–20 m. Myers and Smith (unpublished data) also noted that the length of Dipper territories in the Moorhouse National Nature Reserve was not correlated with river width; territories did not shorten downstream as width increased. By contrast, Sokolowski (1964) found that on streams in the Tartra Mountains in Poland, territories were longer the narrower the stream. Possibly the narrow Polish streams afforded few riffles or for some other reason such as water quality, may have been less favourable to Dippers.

Permanence of Watercourse

Dippers may breed on streams flowing over limestone, which contain water in the spring but generally dry up by mid-summer. Two to three pairs regularly breed on the lower Mounton Brook in south-east Wales but by July this part of the brook runs dry and the adults and young must move elsewhere. Downstream drift of invertebrates during the winter and early spring must be adequate to replenish the food supply for the following year's breeding birds.

Robson (1956) found more Dippers on sandstone areas than on limestone in north Westmoreland, and tentatively linked this difference to more aquatic invertebrates in the former areas. Another explanation may be the tendency for limestone streams to dry up in the summer, or for differences in gradient, water depth or area of riffle between the two sets of rivers.

In Morocco and throughout the Mediterranean, temporary rivers are often physiographically unsuitable for Dippers; their beds may be unstable and their aquatic invertebrate faunas consequently limited by comparison with permanent mountain rivers. Dippers seem to be largely confined to the latter rivers, which

Dipper C. c. minor *river in the Atlas Mountains in Morocco*

are unfortunately becoming scarcer due to modification by man (see Chapter 10).

Nest and Roost Sites: Their Influence on the Abundance of Dippers

On most upland streams it is unlikely that nest or roost sites could limit breeding Dippers. The very nature of the stream, with rocky banks and cliffs, overhanging and fallen trees and branches, means a plethora of possible nest and roost sites. It is possible though on lower reaches where banks may be low and riverside trees absent, or on canalised or otherwise man-modified stretches, that bridges and walls are important in permitting the occurrence and high abundances of breeding Dippers. S. Walker (pers. comm.) believed nest sites (bridges) to be a limiting factor on the R. Theme in Worcestershire, where Dippers are at the eastern edge of their range and where banks are low and trees largely absent. In Germany, Rockenbauch (1985a) also believed that bridges and other riverside man-made structures were essential if high Dipper abundances were to be maintained on the river system that he had studied. Price and Bock (1983) noted that nest sites may also limit populations of the American Dipper.

Other Dipper workers have found that on some rivers where good nest sites are lacking, the provision of nest-boxes can increase the Dipper population. Staedtler and Bremshey (1988) recorded a population increase from 1 pair of Dippers to 14 pairs on part of a river in the Schwerte/Ruhr region of Germany after they had provided nest-boxes. In the Rheingau region, Kaiser (1988) claimed that population increases were due to the provision of 200 nest-boxes, but he accorded this, not to a previous absence of natural sites, but to higher breeding success because of the nest-boxes providing safer nest sites.

Food Supply: Quality, Quantity and Availability

The breeding abundance of Dippers is as described above, partly determined by the *availability* of food which in turn depends on certain physiographic features i.e. a steep gradient providing areas of rocks, riffles, pools and shallow water. Assuming that these features exist, the abundance and type of available invertebrates on a stream bed may have a marked effect on the number of Dipper pairs on that watercourse as noted by Da Prato and Langslow (1976) on the south Esk river system in Scotland.

In 1982 we found marked differences in the breeding abundances of Dippers on different tributaries of the River Wye, despite these tributaries being physiographically rather similar (Fig. 6). Dippers were most numerous on streams in the south-east and central part of the study area, such as the R. Monnow, Grwyne Fawr and R. Edw (Figs. 6 and 8), which overlie limestone or relatively calcareous Old Red Sandstone rocks (total hardness 100–250 mg

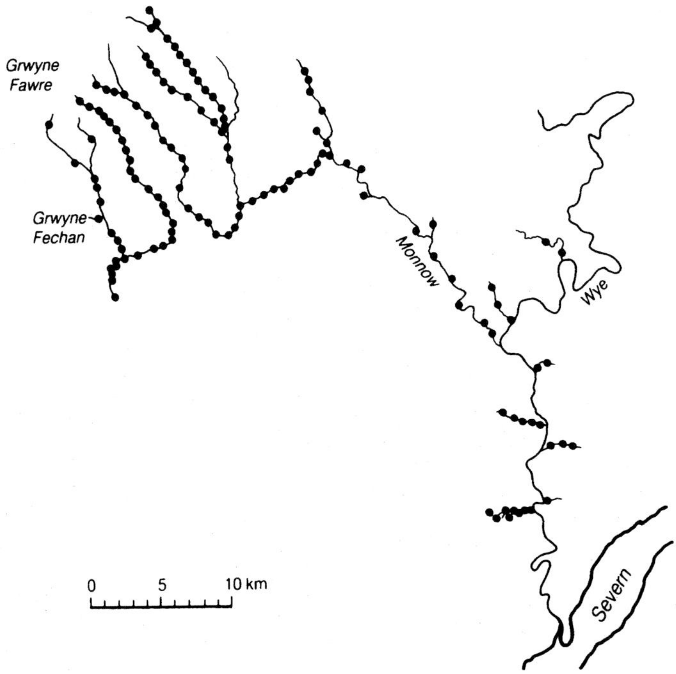

Figure 8. The distribution of breeding Dippers on the good quality rivers in the lower Wye catchment and adjacent Grwyne Fawr.

NOTES: Closed circles denote Dipper territories held regularly, i.e. in three or more years between 1977 and 1993.

$CaCO_3$ per litre), and scarcest on streams such as the upper Wye and the R. Irfon in the west and north, which are underlain by base-poor slates and shales of the Ordovician and Silurian eras. On these latter watercourses Dippers were not found for some distance downriver from their sources, and contrary to the pattern elsewhere, Dipper abundances were greater further downriver (Fig. 9). From kick-sampling invertebrates on these Wye tributaries in 1982 it was apparent that Dippers were most abundant on streams which had the highest abundances of some large invertebrates, and were scarce or absent on those watercourses where invertebrate faunas were impoverished (Ormerod, Boilstone and Tyler, 1985). Dipper breeding abundance was strongly correlated, particularly with the relative abundance of larval caddis flies (Trichoptera) and nymphal mayflies (Ephemeroptera), as well as nymphal stoneflies (Plecoptera) (Fig. 10). In Chapter 4 we show that caddis larvae and mayfly nymphs are of particular importance as prey to breeding adult and nestling Dippers, whereas stonefly nymphs comprise only a small proportion of the diet of adult or nestling Dippers on good quality streams. Because there was a significant correlation between the abundance of stoneflies and gradient, we suggested that the abundance of Dippers and of plecopterans was spuriously related through this

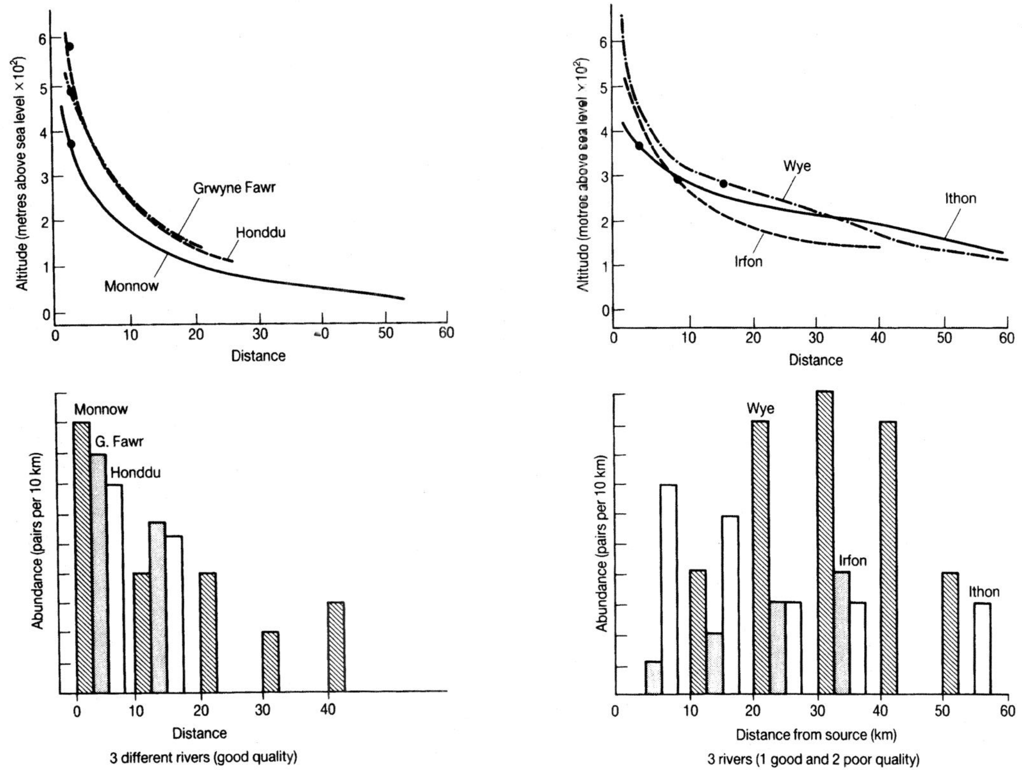

Figure 9. The distribution of breeding Dippers on five rivers in the Wye catchment and on the adjacent Grwyne Fawr.

NOTES: The upper figure shows the altitudinal profile and circles denote the position of the nearest occupied territory to the source. The histogram represents the abundance of breeding Dippers in each 10 km stretch (Ormerod, Tyler and Lewis, 1985).

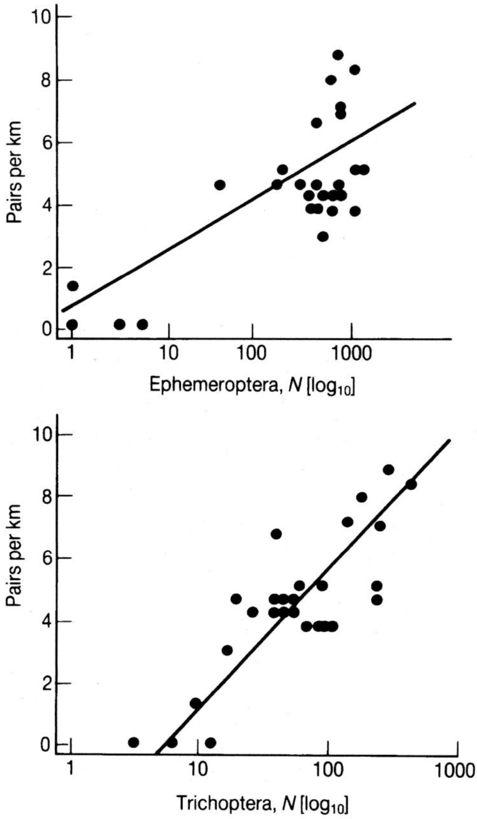

Figure 10. Relationships between the breeding abundance of Dippers (pairs per 10 km) in the Wye catchment and the abundance of (top) nymphal mayflies and (bottom) larval caddis.

NOTES: N = numbers per 2 × 3-minute kick samples; log scale (Ormerod, Tyler and Lewis, 1985).

third factor. There was no apparent correlation between gradient and the abundance of larval caddis.

The presence of different species of benthic invertebrates and of some fish is strongly correlated with water quality, particularly factors such as pH, calcium and total hardness and the concentration of aluminium. Thus many genera of caddis and mayfly and other invertebrates such as freshwater shrimps *Gammarus* and molluscs are known to be scarce or absent from rivers with a pH below 5.4–5.7 and at aluminium concentrations of over 1.0–1.5 mg/l. Those Wye tributaries which in the 1982 survey had few species and low abundances of invertebrates, were all relatively acidic (mean pH 6.0–6.5, but minima of 4.5–5.0), and the breeding abundance of Dippers on the upper reaches of all the

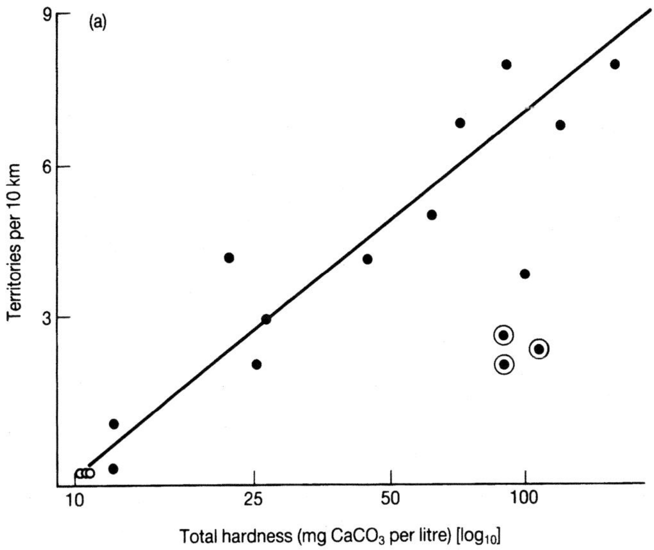

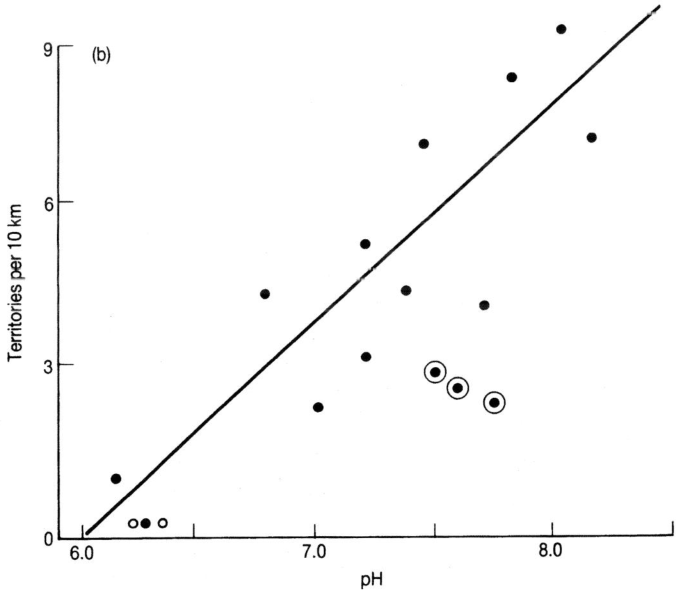

Figure 11. The abundance of breeding Dippers on the upper reaches (nominally 10 km) of 16 tributaries of the Wye and the adjacent Grwyne Fawr in relation to: (a) mean long-term pH, and (b) total hardness.

Wye tributaries correlated with both mean pH and calcium and total hardness (Fig. 11).

Subsequently, to assess the distribution of Dippers with respect to acidity over a restricted range of calcium and hardness concentrations, the RSPB in 1985 looked at 74 stretches of 1 km on soft-water streams (7–25 mg $CaCO_3$/l) throughout mid- and north Wales (Fig. 12). Dippers were generally scarce on these streams (Ormerod *et al.*, 1986), birds being found at only 21 (28.3%) of the sites. Paired comparisons of environmental variables and subsequent multiple discriminant analysis using ten key variables showed that sites without Dippers in this survey had significantly lower mean pH and higher concentrations of filterable aluminium (>0.1 mg/l) than those where Dippers were present (Fig. 13), whilst total hardnesses were not different. Their presence was also strongly correlated with the abundance of both caddis and mayflies (Fig. 13) and sites

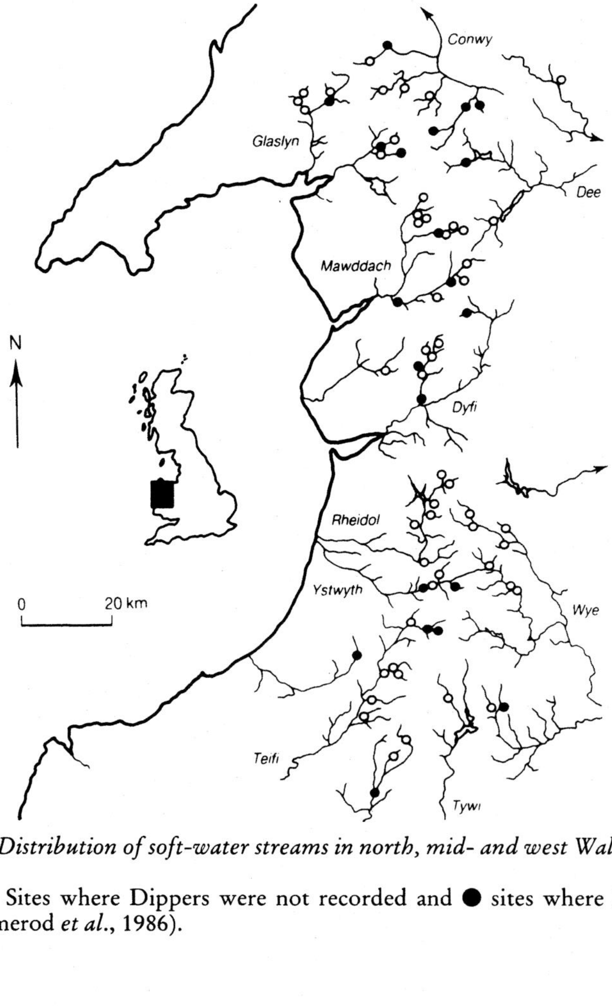

Figure 12. Distribution of soft-water streams in north, mid- and west Wales surveyed in 1984.

NOTES: ○ Sites where Dippers were not recorded and ● sites where Dippers were present (Ormerod *et al.*, 1986).

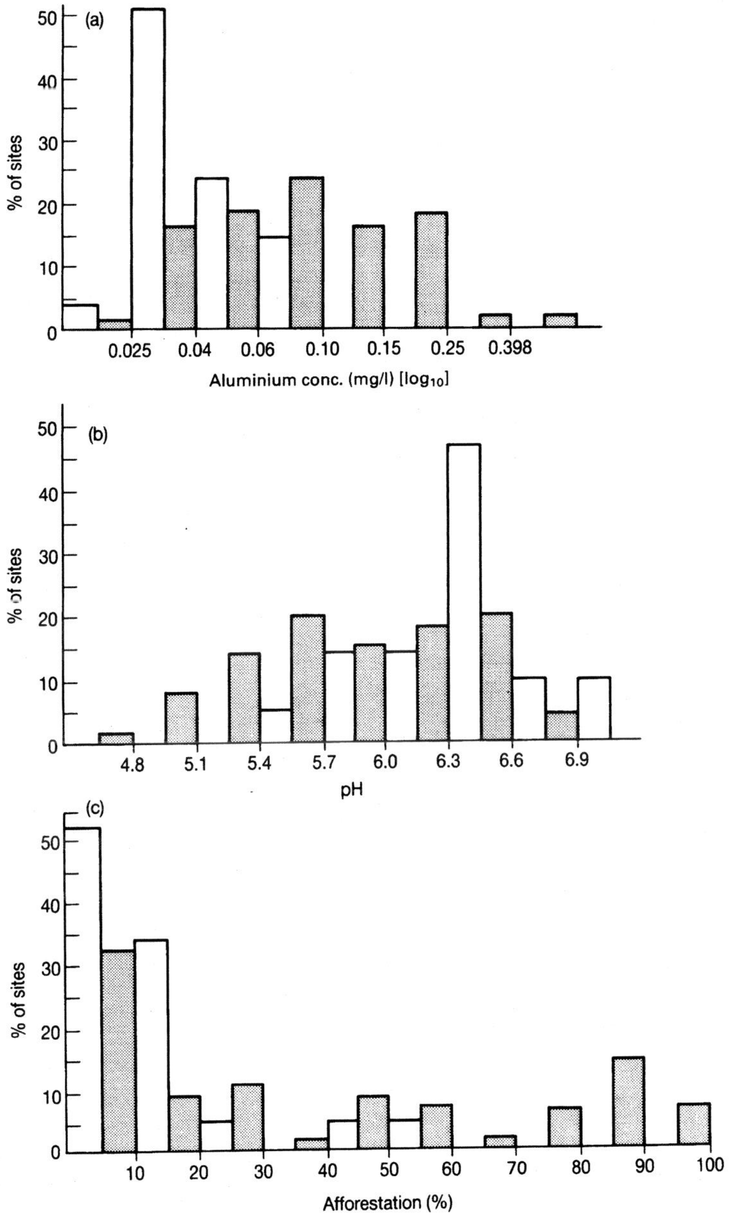

Figure 13. The factors on soft-water sites in 1985 which determined whether Dippers were present (open bars) or absent (shaded bars).

NOTES: (a) Aluminium concentrations in the water; (b) mean pH; and (c) percentage of afforestation in the catchment (Ormerod *et al.*, 1986).

without Dippers had fewer broadleaved trees on their banks and more conifer afforestation on their catchments than where Dippers were present (Fig. 13).

We believe that increased acidity and aluminium concentrations influence the abundance of Dippers largely through affecting the composition and abundance of preferred invertebrate prey. Additional evidence for an impact through the food supply comes from the longer territories defended by pairs of Dippers breeding at low density on acidic streams compared with birds breeding elsewhere (Fig. 14). We suggest, too, that the scarcity of calcium in acidic streams may also limit clutch size and growth of Dipper nestlings, as well as posing additional stress on adult females prior to egg-laying, calcium having to be mobilised from bones for eggshell formation (Ormerod *et al.*, 1991).

The impact on Dippers of increased surface-water acidification due to high sulphur emissions and the exacerbating effect of extensive conifer forests in the upland catchments is later discussed (Chapter 10), but we note here that there is evidence from historical data for one tributary of the R. Wye, the upper Irfon, that acidity has notably increased since the early 1960s, with a decline of 1.7 units in pH measured in winters between the 1960s and 1984, during which time the Dipper population has fallen from nine pairs in the upper 10 km to just one to two pairs (Fig. 15). Acidification of watercourses elsewhere may be implicated in some of the changes in the distribution of Dippers between 1968–72 and 1988–91 (Fig. 16).

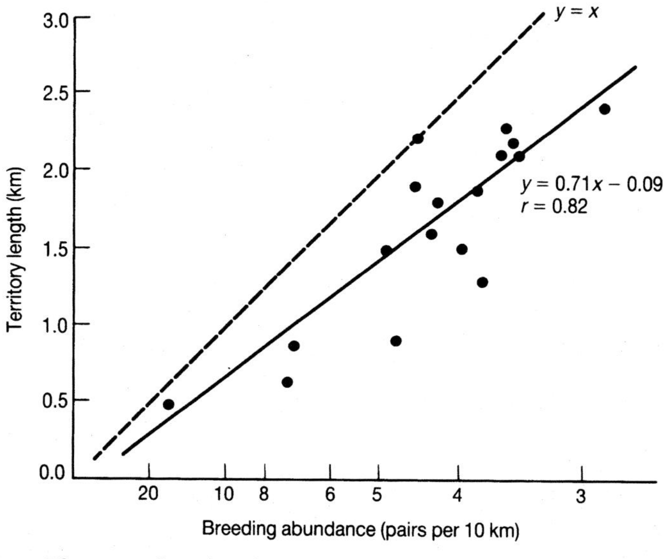

Figure 14. The mean lengths of territories held by breeding Dippers plotted against abundance (shown as its reciprocal).

NOTES: The dashed line shows the maximum possible territory length at any abundance (Ormerod and Tyler, 1987b).

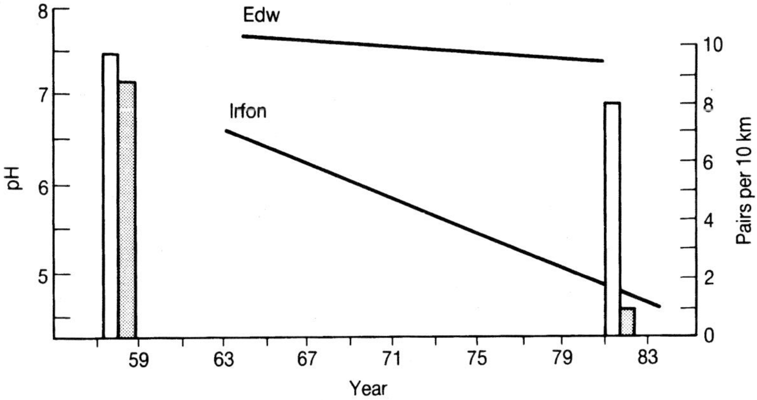

Figure 15. The breeding abundance of Dippers along the River Edw (open bars) and River Irfon (shaded bars) in the mid-1950s and 1982.

NOTES: The regression lines show trends in spot-metred winter pH (Ormerod, Tyler and Lewis, 1985).

Acidity is only one form of pollution, although it does affect thousands of kilometres of upland streams in Wales alone, as well as watercourses in Scotland, parts of Europe and elsewhere where there are base-poor soils, i.e. which have poor buffering capacity. Other forms of pollution, whether from agriculture or industry, may also affect sensitive invertebrates such as mayfly nymphs, through reducing the oxygen in the water, and thereby affect the abundance of Dippers on those watercourses. Serle and Bryson (1935) and Cowper (1973) found that pollution in the form of effluent from paper mills, coal dust and sewage all reduced the numbers of Dippers on the R. Esk in Scotland, and Cole (1990) on the Derbyshire Wye also recorded fewer Dippers on sections of the river affected by organic pollution. There is some evidence, too, from south-west Wales that Dippers are scarce on streams suffering from chronic silage and slurry pollution (Tyler, 1990). Dippers were also formerly absent from many rivers in the industrial south Wales valleys but with the decline in coal mining and other industries, they have recolonised many of these now cleaner rivers.

Vegetation along a stream can also influence the availability and abundance of prey and hence the abundance of Dippers. For example, broadleaved trees overhanging a stream contribute leaves and other organic matter which is broken down by shredding and grazing invertebrates, themselves prey for Dippers. Trees, notably oaks, also contribute many terrestrial invertebrates to the river: caterpillars, other insects and spiders fall from the tree into the water, where they may be eaten by other invertebrates, fish or even by Dippers. Conifers may, however, have a deleterious effect. The dense shade caused by spruce trees and other conifers overhanging a watercourse, and possibly, the low water temperatures during the summer in streams flowing through conifers, result in an

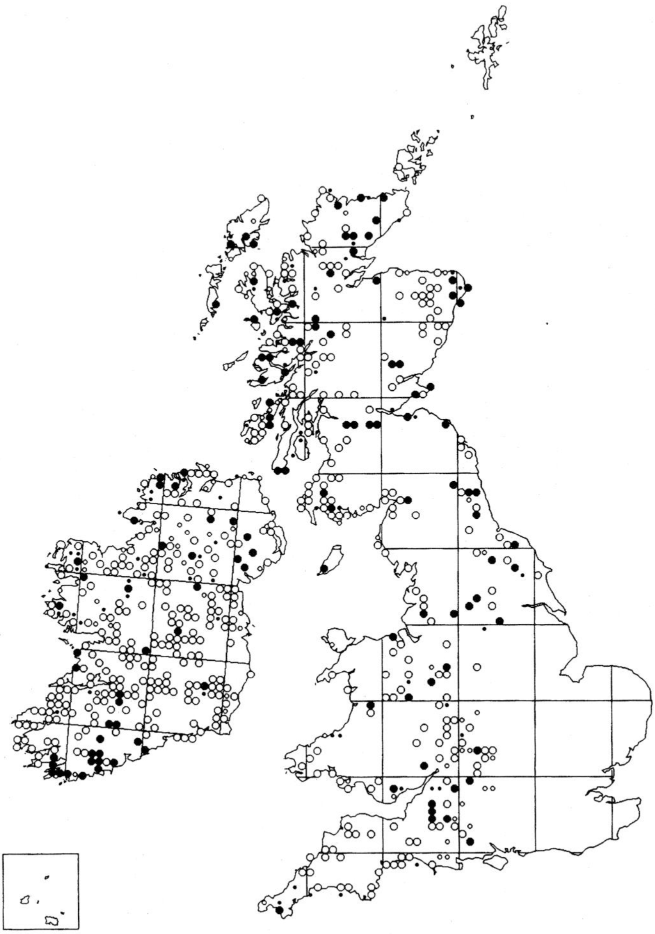

Figure 16. Changes in distribution of breeding Dippers in Britain and Ireland between 1968–72 and 1988–91.

Small open dot = possible breeding in 1968–72, not recorded in 1988–91; large open dot = probable or confirmed breeding in 1968–72, not recorded in 1988–91; small filled dot = not recorded in 1968–72 but present, although no evidence of breeding, in 1988–91; large filled dot = not recorded in 1968–72, but evidence of breeding in 1988–91.

From *The New Atlas of Breeding Birds* (Gibbons, Reid and Chapman, 1993) with permission from the British Trust for Ornithology.

impoverished benthic invertebrate and fish community. We know from surveys in Wales that Dippers are scarce on such streams irrespective of water quality. Other factors which reduce water temperatures or cause reduced invertebrate abundances have a similar effect. Thus where cold water is released from the bottom layers of a reservoir, invertebrates and fish are adversely affected for several kilometres downriver. Rivers immediately below impoundments may also have reduced invertebrate faunas because of a flushing effect from the released water. (Conversely, sometimes they have increased abundances of filter-feeding invertebrates because of the planktonic plants and animals washed out of the lake.) There are confounding factors too such as the water from bottom releases from reservoirs being likely to be acidic or have elevated metal concentrations.

In summary, Dippers prefer permanent rivers, with a steep gradient and with clear, fast-flowing well-oxygenated, unpolluted water with plentiful benthic invertebrates (particularly caddis larvae and mayfly nymphs) for food; mid-stream rocks as perches, areas of riffles, pools and shallow water in which they can forage, and safe nest and roost sites in banks, on cliffs or in or on walls and under bridges.

Where all these requirements are met, Dippers breed at high abundances, with contiguous territories of 500 m in length or less. Rivers with a shallow gradient, with deep, slow-moving or turbid water, with a silty or sandy substrate, with poor water quality and impoverished invertebrate faunas, or with a lack of suitable nest or roost sites will support few if any Dippers. Bankside cover is not essential. Birds will breed at high abundances on rivers lined with alders or fringed with deciduous woodland, but may also breed on streams flowing through open moorland, agricultural land, towns or cities. Only where conifers overhang a river, causing dense shade in the spring and summer and hence reduced invertebrate faunas, are birds absent or scarce. Dippers are also scarce on acidic streams, where there is a limited supply of calcium for egg formation and growth and where there is a reduced invertebrate fauna, particularly caddis larvae and mayfly nymphs, both groups of which are important Dipper prey.

DISTRIBUTION OF OTHER SPECIES OF DIPPER

Other species of dipper have similar requirements, occurring on upland rivers down the western side of the North American continent, in the mountains of the Andes and in the Himalayas and other mountain ranges in Asia, China and Japan, but being absent from rivers in lowland areas other than as occasional visitors or stragglers in the winter.

Few studies have been made of factors which influence the breeding abundance of these other species of dipper but the sparse data available suggest that similar factors influence territory length and breeding abundance of American and Brown Dippers as influence the Eurasian species. Thus in Colorado, Price and Bock (1983) found that food availability correlated positively with the number of American Dippers breeding along one river. Asami and Haga (1983) who surveyed Brown Dippers on the upper reaches of the Tokachi River system near Hokkaido in Japan, showed a correlation between the length of their home ranges (and hence overall breeding abundance) and various environmental factors, notably the nature of the stream bed and the composition and relative abundance of invertebrate prey. Interestingly, Asami and Haga claimed that in the early spring pairs first established home ranges; these were on average about 900 m long, with home ranges of adjacent pairs often overlapping. Territories were then established within the home ranges, occupying around 80% of these. Sunquist (1976), however, who carried out a detailed survey of Brown Dippers on 4.4 km of a narrow (4–7 m wide) torrential stream in the Chitral District of Pakistan, recorded birds holding contiguous territories. His average territory size was 425 m (range 315–562 m) with territories contiguous and occupied by monogamous pairs.

Our observations in Nepal also indicate that the same physiographical features, as well as clear unpolluted water with an abundance of favoured insect prey, determine the presence and abundance of Brown Dippers, as they do Welsh Dippers (Tyler and Ormerod, 1993). Thus we found pairs of Brown Dippers to be frequent in late March on the Langtang Khola, a clear, rocky mountain river with a steep gradient, at an altitude of *c.* 2000–3500 m. Here there was a diverse aquatic invertebrate community dominated by such families as mayflies, caddis and blackflies. On the Likhu Khola, in an intensively cultivated area of the Middle Hills of Nepal, Brown Dippers were generally scarce even on the rocky, steep tributaries. This we attributed to the often turbid water and deep silt deposits on the stream beds, and the greatly reduced flows — the result of run-off of soil from the cultivated slopes and diversion of water to irrigate the surrounding terraces. These watercourses generally had few caddis and mayflies. On the Bagmati River in the Kathmandu Valley, Brown Dippers occurred on wooded, rocky stretches above the city but were absent from a rocky gorge below the city, undoubtedly because the river here was grossly polluted and hence had no suitable prey for these birds.

Limited data collected in October 1993 on Rufous-throated Dippers suggest that birds prefer stable sections of upland river in cloud forest where there are rock cliffs or boulders to provide nest sites and shelter, and where invertebrate prey is plentiful and readily accessible. Only one bird was seen on a grossly enriched but physiographically suitable 7-km section of river in the Aconquija range in Tucuman province of Argentina, whereas pairs were present on a clean tributary from its confluence with the main river and up for the 3 km surveyed at *c.* 1 km intervals. The dense growth of filamentous algae in the main river had probably reduced the mayfly and caddis fly fauna. On another river in this range birds were absent from a 4-km section with an unstable bed of rounded boulders, and where there was an absence of rocks or cliffs, but in a narrower gorge above with cliffs and numerous waterfalls, three pairs were found in under 3 km (Tyler, 1994).

Dipper with head underwater looking for prey

CHAPTER 4

Food and feeding

Upland rivers are amongst the most dramatic landscape features on earth. Seeping from mountain glaciers, carving through ancient gorges and cascading over boulders and sheer faces of waterfalls, they both follow the land and shape it. They are, though, far more than simple physical phenomena. Beneath their surface, entrained in the flow or anchored to their beds, micro- and macroscopic plants turn energy into living biomass. Leaves and litter fall from the surrounding land into the stream and are processed and assimilated by bacteria and fungi. Many insects and other invertebrates occupy a myriad of habitats in or on the river bed, feeding on an abundance and variety of tissue. Here, mayflies graze algae from the surface of the stones and larger plants. Caddis larvae spin their silken nets to capture particles carried in the flow. Other caddis and stoneflies shred leaves and litter. Predators stalk or wait to prey on other invertebrates. Finally, fish and amphibians exploit both the plants and animals in the river and its surroundings. Clearly, here is a rich and productive source of food for non-aquatic organisms which have evolved the special traits that allow them access to it. It is this ecosystem whose production fuels the five species of dipper, on which they are almost wholly dependent, and towards which their adaptations are directed.

ENERGY REQUIREMENTS

Averaged throughout the year, a Dipper in a temperate region such as Wales needs about 173–213 kilojoules (kJ; 41–51 kcal) of energy daily. This satisfies its needs for activities such as foraging, flight, diving, roosting, keeping warm and basal metabolism. Males being slightly larger, require about 15% more energy than females, though female requirements rise slightly during reproduction due to the demands of egg production and incubation. Both sexes must also replace their feathers annually, adding roughly 16–18 kJ per day to energy costs over this period. Thus, the annual energy requirements of a pair are around 153,000–158,000 kJ. The adults need to harvest a further 10,000–20,000 kJ from their riverine territory to feed one or two broods of young. These total requirements are equivalent to about 8–9 kilograms (kg), as dry weight, of food per year. However, since insectivorous birds actually digest and assimilate only about 70% of the food they eat, the amount of food which a pair of Dippers and their brood(s) need each year is likely to be nearer to 11–13 kg dry weight. It is the satisfaction of these needs which gives rise to the array of foraging strategies used by Dippers.

FORAGING BEHAVIOUR

As with any bird, finding food to satisfy energetic needs is crucial to the Dipper, but a range of factors constrain the time and opportunities available for foraging. Some constraints are biological, such as the need to watch for predators, defend territories, guard their mates, digest their food, rear their young and maintain their feathers. Others are abiotic as, for example, the onset of darkness or adverse conditions such as floods or freezing over of rivers. Foraging may also be restricted by food scarcity. Thus, strategies governing how, when and where to feed, and what food to take, will have evolved to enable Dippers to optimise their feeding activity in the wake of all these constraints. At the same time, the foraging success of young birds may be influenced by their incomplete knowledge of the nature and distribution of food, and hence by their need to learn.

Averaged throughout the year, Dippers spend 45–55% of the daylight hours foraging, with remarkably similar patterns apparent in Germany, Scotland and Wales (Fig. 17). On acidic streams, where food is limited, foraging time increases significantly by comparison with food-rich circumneutral streams (55% cf. 38%). Most of the remaining time is spent resting (digesting, preening, etc.), with other activities such as flying and singing taking only a small amount of the day (Fig. 17). There are marked changes through the year, though, with similar results again coming from different studies (Fig. 18). During winter, foraging

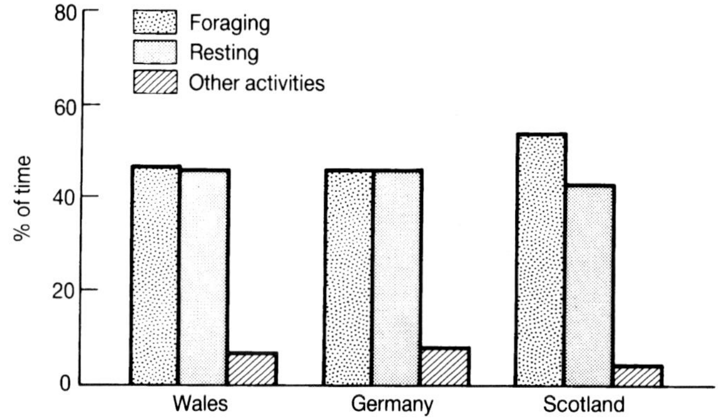

Figure 17. Time activity budgets of Dippers in Wales (our data), Germany (Spitznagel, 1988) and Scotland (Bryant and Tatner, 1988).

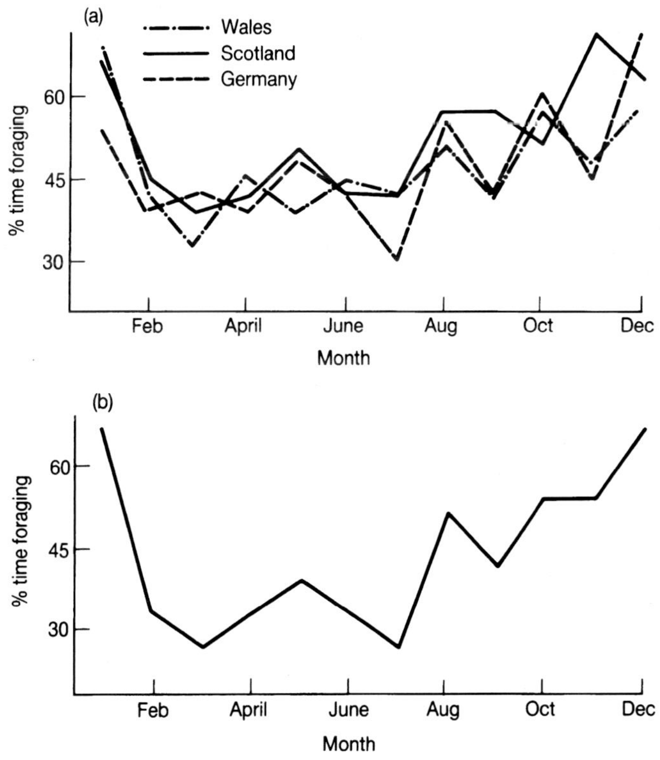

Figure 18. (a) Seasonal changes in time spent foraging in Wales (our data), Scotland (Bryant and Tatner, 1988) and Germany (Spitznagel, 1988). (b) Average time spent foraging through the year (all sites combined).

takes up over 60% of the active day, while in May, provisioning nestlings with food causes a moderate increase in foraging. By contrast, during the moult in July the amount of time given to foraging decreases to less than 30–40% of the active day, so that much of the daily energy requirement must be gleaned during this time.

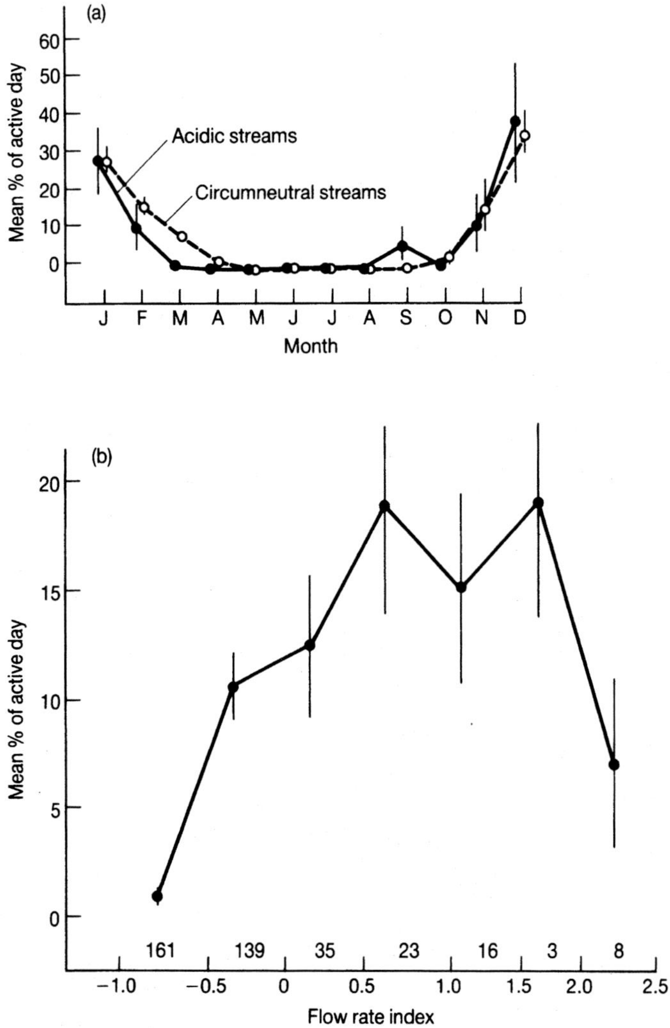

Figure 19. (a) Changes in diving activity of the Dipper through the year. (b) Diving activity of the Dipper in relation to stream flow.

NOTE: The flow rate index shows changes in flow relative to the annual mean. Sample sizes at each flow are shown.

The Dipper's foraging activity is essentially divided into diving, swimming, wading at the river margins turning stones and leaves, and gleaning stone surfaces. Activities such as fly-catching and gleaning plants also occur, but fewer than 1.5% of foraging bouts involve these methods (Spitznagel, 1988). The major foraging techniques are diving (*c.* 56% of bouts) and wade-prying (37% of bouts). The relative frequencies of these two methods change substantially during the year, with diving most common during the period of November to March (Fig. 19a). In part, this probably reflects the strong effect of increasing river flows which increase the frequency of diving (Fig. 19b). At the highest flows however, diving begins to decline, probably because rivers in spate become too turbid and turbulent for effective foraging. Dippers at these times can switch to feeding in smaller streams (da Prato, 1981), perhaps concentrating on stone-surface insects such as blackfly larvae and *Baetis* mayflies. These insects are common in the diet during the winter (see Fig. 27). Terrestrial prey such as scarabid and elaterid beetles have also occurred in Dipper faecal samples collected during high flows in Wales, though it is not known whether these were picked from terrestrial surfaces or whether the beetles had fallen into the river.

WHERE TO SEARCH

Rather than being uniform in character, rivers are a mix of different habitats: riffles, rock surfaces, pools, margins and vegetation. Macroinvertebrates and fish show differences in distribution between these habitats, while physical aspects such as current speed and turbulence create further heterogeneity. Both Thut (1970) and Spitznagel (1988) have, respectively, assessed how foraging by American and White-throated Dippers is spread over different habitats by looking at the microdistribution of their prey. Their views coincided, indicating that both species predominantly used backwaters, pools and river margins, 62–95% of prey being from these habitats in Spitznagel's German study. In Wales, the limnephilid caddis larvae often eaten by Dippers are concentrated in the river margins (Ormerod 1988) although blackfly larvae occur mostly in riffles. Direct observations by Chambers (1980) confirm diving to be commoner in the margins than in the mid-reaches, with wade-prying also likely to be easier here than in fast riffles.

BROOD PROVISIONING

Provisioning a dependent brood by transferring food to the nest imposes special constraints on the foraging behaviour of adult birds. Flying to and fro many times a day is liable to be costly in terms of time and energy. They may also

be vulnerable to predators as they leave their normal cover and repeatedly pursue a path to the nest. At the same time, the growing nestlings have increasing energy demands which the adults must satisfy to ensure their offspring have the best chances of survival, and to prevent them calling so much that they attract predators. Early on, however, the nestlings must also be kept warm, and the rate at which they can ingest food may be limited by the size of their gape and gut. Strategies of provisioning the brood are thus liable to have been finely honed by evolution to meet all these crucial needs in an economic way. In the Dipper, provisioning the nest is subject to a further, particular, constraint because of the linear shape of the territory. A Dipper territory of say 5000 m^2 on a river 8 m wide will be around 625 m long. A nest at its middle could then involve flights of up to 312 m. Another species of bird with a territory of similar area, but of circular shape, will never be more than about 40 m from a centrally placed nest.

Early in the brood period each Dipper nestling is fed on average 4.3 times per hour, both male and female sharing feeding roughly in equal proportions (Fig. 20). Load size at seven days is around 36 mg dry weight, and small prey are taken, probably because of the small nestling gape and gut. There may be larger numbers of prey in these earlier loads (10–20 mayflies, 6–7 small caddis, or a mix) than in the loads given to older nestlings. Marked changes then occur as the

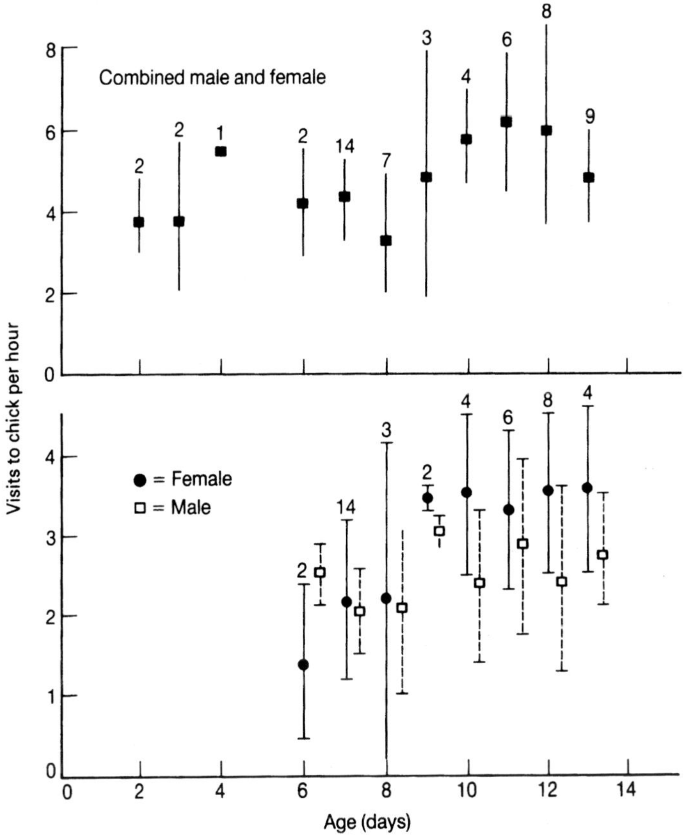

Figure 20. Brood provisioning by adult Dippers in relation to nestling age.
NOTE: The vertical bars are standard deviations headed by sample size.

nestlings grow and their energy demands increase. The load size increases to around 73 mg dry weight by Day 11, with increasingly large items being delivered (see below). Almost invariably, adults carry several prey to the nest at once, such as 4–9 caddis larvae. Fish and sometimes dragonfly nymphs such as *Cordulegaster boltoni* are fed as single items. The rate of feeding visits also increases to 6.2 visits per chick per hour, so that a pair feeding a brood of five will each need to collect 30–60 large prey items per hour. The change in provisioning rate does not, however, appear to be a simple function of age (Fig. 20). Instead, the average provisioning rate shows a rough step change at around Day 8 or 9, particularly by the female as she is freed from the demands of brooding. By this time, her contribution of about 55% of the feeding is slightly greater than the male's. This contribution changes markedly in instances of polygyny, where the female performs over 60–70% of the feeding visits. In the few instances of widowhood we have observed, the female has increased her feeding rate, presumably at considerable risk and cost to her own welfare. Her total visit rate does not match that where both parents are present and widows' broods are usually very small. With these smaller broods, individual chicks can receive as many feeds as those where both parents are alive.

A particularly impressive instance of brood provisioning was recorded by Alder (1963) during heavy flooding in Northumberland. Rising water had cut off a nest with a brood in an air pocket beneath a bank, but both parents kept up their feeding visits which involved diving to reach the nest. Diving, we should remember, is one of the most energetically costly activities that a Dipper ever undertakes.

Other factors appear to affect the rate of feeding visits. For example, the number of feeds per nestling decreases in larger broods, or falls by about two-thirds as the breeding season progresses. Nestlings reared on acidic streams, where food is limited, also receive about two-thirds of the feeding visits of birds on food-rich streams. In none of these instances do we know if load size changes to accommodate the altered rate of feeding, though all these factors reduce growth and survival in ways which could be consistent with a reduction in food provisioning.

Adults also alter their rate of brood provisioning through the day. Marked peaks in feeding occur shortly after dawn and before dusk, with a reduction to a minimum around midday (Fig. 21). We do not know whether this is to allow the

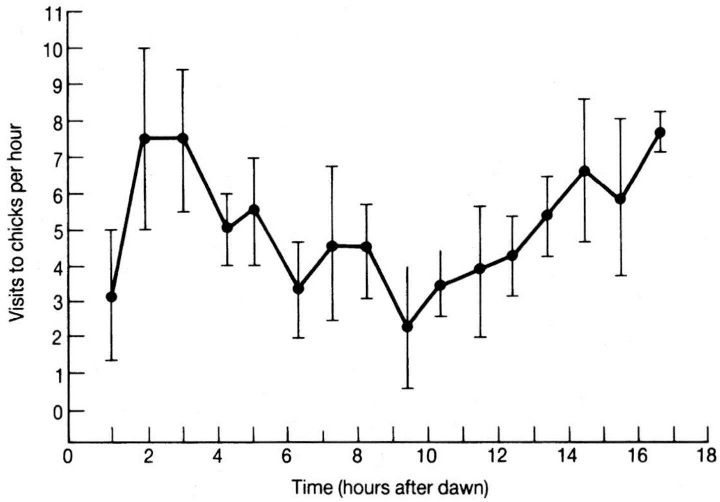

Figure 21. Diurnal changes in brood provisioning by adult Dippers on circumneutral streams.

birds to rest or preen, or whether it reflects some other constraint. Hunting success may, for example, decline as river temperatures rise and prey become more active, or adults may find it difficult to keep cool during vigorous activity in the heat of the day. The pattern may also reflect the physiology of the chicks, for example, as a result of energy demands overnight, or limits to their rate of digestion during feeding.

Other Circadian Patterns

Initial evidence that there were marked diurnal patterns in Dipper foraging at other times of the year came indirectly from circadian changes in weight. After losing weight while roosting overnight, a marked peak in weight occurs within one or two hours of dawn as a result of intense foraging activity during the first hour of daylight (Fig. 22a). Daily activity budgets have further shown this to be the case, with foraging declining through the day while resting increases (Fig. 22b). The same pattern is less marked, however, on acidic streams where food is most scarce. Figure 22c shows diving activity through the day.

FOOD

Methods of Study

The earliest methods used in assessing the food of dippers involved analyses of gut contents of dead birds, often shot specifically for the purpose. Most studies now use less extreme methods such as field observation, photography and, most commonly, analyses of regurgitated pellets and faeces. In dippers, the last two methods have been easy by comparison with many other insectivorous birds which prey opportunistically on a wide variety of terrestrial families. Dippers have access to a relatively restricted range of prey whose identification is facilitated by their well-known taxonomy and the persistence through digestion of structures such as insect mouthparts and fish bones. These can be counted accurately and can also be often used to reconstruct the original size or weight of the prey.

Studies of the Dipper in different parts of its range have shown remarkably similar diets and generally similar changes in diet through the year. The diet is dominated overwhelmingly by prey of aquatic origin, with aquatic stages of insects being numerically dominant along with molluscs, crustaceans and fish (Figs. 23–27). Often some less frequent prey, such as fish or certain caddis larvae, make a disproportionately large contribution to the weight of the diet because of their larger size. These groups thus dominate the diet by weight for most of the year and in most locations (Table 4), but some interesting numerical changes occur between sites and seasons.

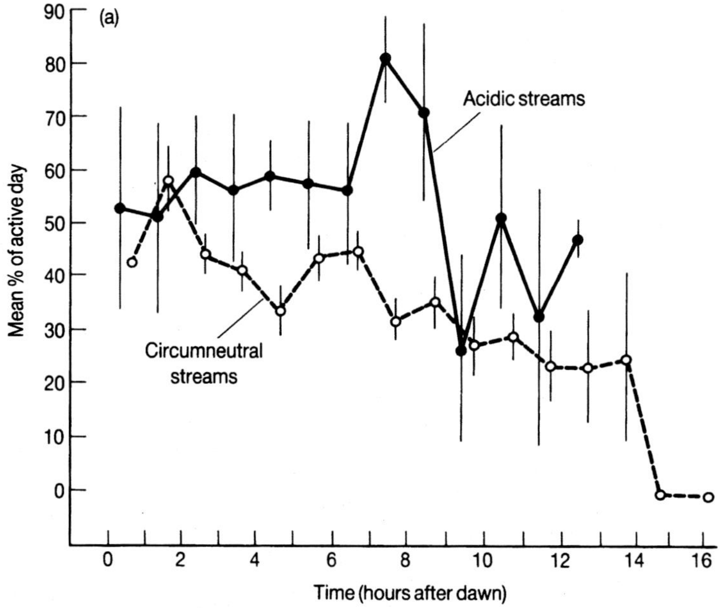
(a)
Acidic streams
Circumneutral streams
Mean % of active day
90
80
70
60
50
40
30
20
10
0
0
2
4
6
8
10
12
14
16
Time (hours after dawn)

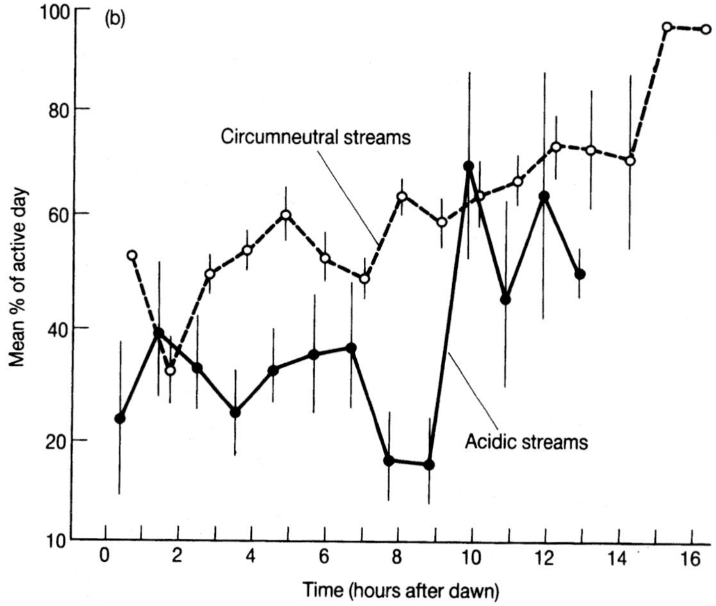
(b)
Circumneutral streams
Acidic streams
Mean % of active day
100
80
60
40
20
10
0
2
4
6
8
10
12
14
16
Time (hours after dawn)

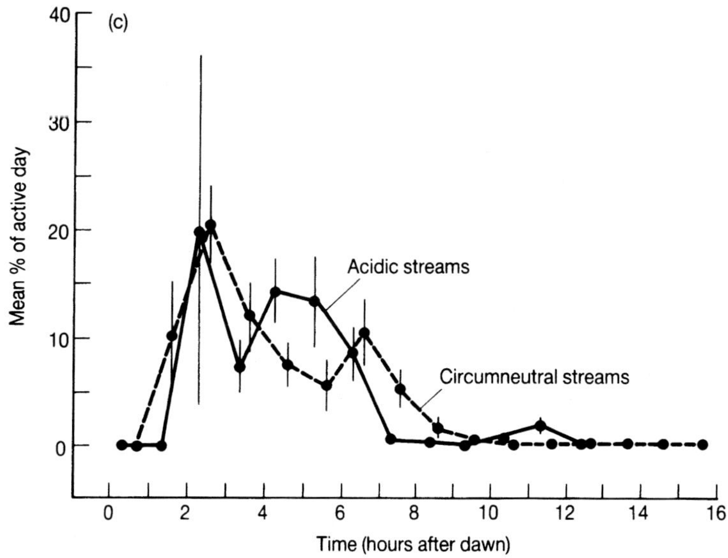

Figure 22. Diurnal changes in the Dipper's activity patterns: (a) foraging, (b) resting and (c) diving.

Diet During the Breeding Season

During the breeding season, site-to-site consistency in the adults' diet is very marked (Fig. 23) with caddis larvae and mayfly nymphs almost completely dominant but stoneflies taken in large numbers on acidic streams. The diet of nestlings changes markedly as the chicks grow. Young nestlings are fed predominantly on *Baetis* mayflies (1–2 mg individual dry weight), with limnephilid and hydropsychid caddis larvae (6–60 mg individual dry weight) becoming increasingly important as the nestlings grow (Fig. 24). Adults take smaller prey for themselves over this period. Similar patterns occur along acidic streams except that stonefly nymphs, instead of mayflies, are fed to young chicks (Fig. 23b). Almost certainly, the selection of large prey increases the efficiency of food delivery to the nest, with prey size an important criterion for adults in deciding which prey to feed to nestlings (Figs. 25 and 26). This selection for larger prey appears to be much stronger between prey types than within them since the size of items fed to chicks and eaten by adults does not always differ markedly within families, nor in a consistent way (Table 5). The exceptions are in instances where a wide range in the size of available items occurs within a family, such as the perlid stoneflies. Here, the larger prey are usually fed to chicks. Other features thought to promote positive selection include the generally slow movement and low degree of sclerotisation of caddis larvae (Spitznagel, 1985c). Interestingly, at

Table 4. The estimated percentage contributions (by weight) to the diets of Dippers on acidic and circumneutral streams in Wales at different times of the year

	Breeding[1]					
	YN	*Adults*	*ON*	*Adults*	*Moult*[2]	*Winter*[3]
CIRCUMNEUTRAL STREAMS						
Molluscs	0	0	0	0	0	6.1
Crustacea	0	0	0	0	0	3.2
Ephemeroptera	35.3	19.6	3.8	17.1	7.3	3.2
Plecoptera	18.8	4.9	0.8	1.7	0.7	1.3
Trichoptera	45.7	75.4	80.9	51.4	69.4	19.3
Diptera	0.2	0.1	0.1	0.8	3.4	3.1
Others	0	0	8.5	1.6	0.1	0.3
Fish	0	0	5.9	27.4	19.1	63.8
ACIDIC STREAMS						
Molluscs	0	0	0	0	—	0
Crustacea	0	0	0	0	—	0
Ephemeroptera	6.1	0	0.2	1.3	—	0
Plecoptera	20.2	33.5	4.2	36.6	—	32.1
Tricoptera	73.6	64.6	95.5	61.3	—	60.1
Diptera	0.1	1.9	0.1	0.8	—	7.7
Others	0	0	0	0	—	0.1
Fish	0	0	0	0	—	0

NOTES:

Breeding is taken as April–June inclusive, Moult as July–September inclusive, and Winter as October–March. No data were available from acidic streams during moult. Data from the breeding season have been combined for nestlings >5 days old (ON) as opposed to those <5 days old (YN).

Note all data were either from Ormerod (1985a)[1], from Smith and Ormerod (1986)[2], or from Ormerod and Tyler (1986)[3].

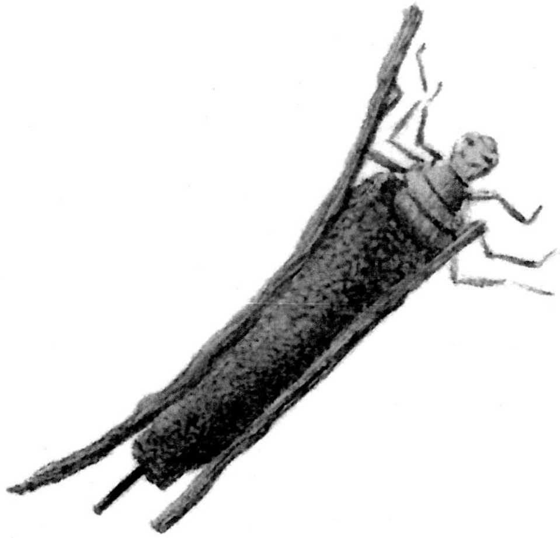

Cased caddis larva

Net-spinning hydropsychid caddis larva

Mayfly nymph Ephemerella

Stonefly nymph Perlodes

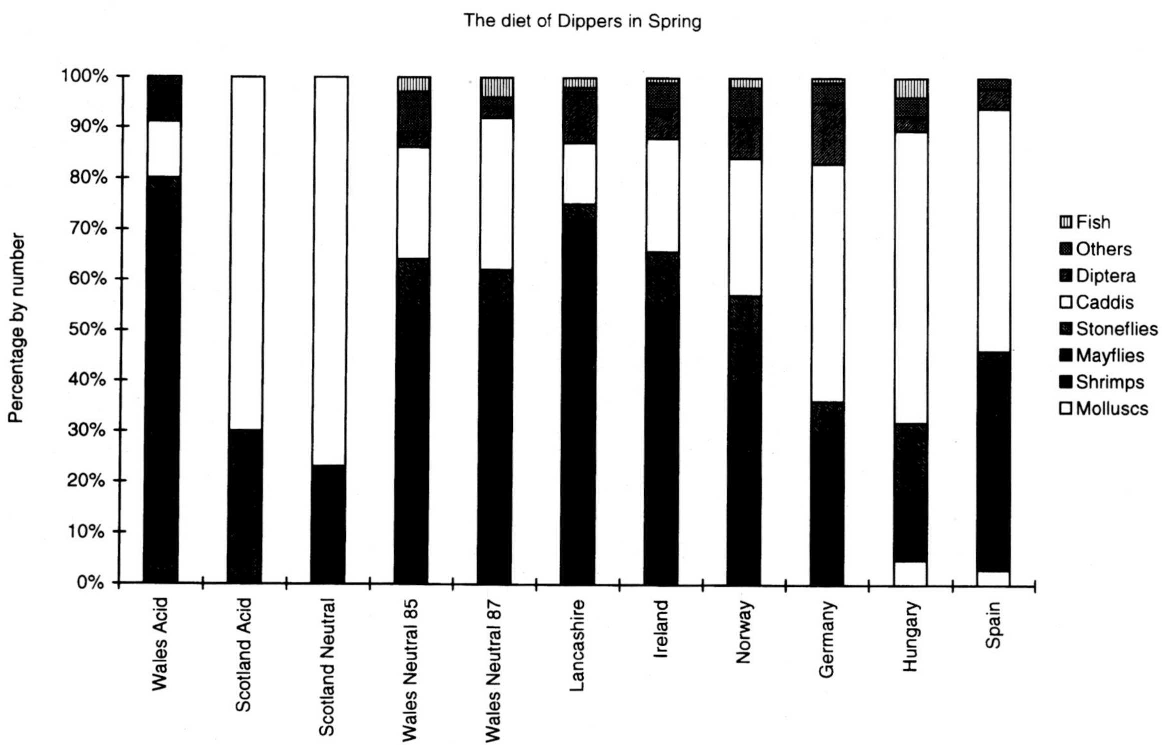

The diet of Dippers in Spring
Percentage by number
100%
90%
80%
70%
60%
50%
40%
30%
20%
10%
0%
Wales Acid
Scotland Acid
Scotland Neutral
Wales Neutral 85
Wales Neutral 87
Lancashire
Ireland
Norway
Germany
Hungary
Spain
Fish
Others
Diptera
Caddis
Stoneflies
Mayflies
Shrimps
Molluscs

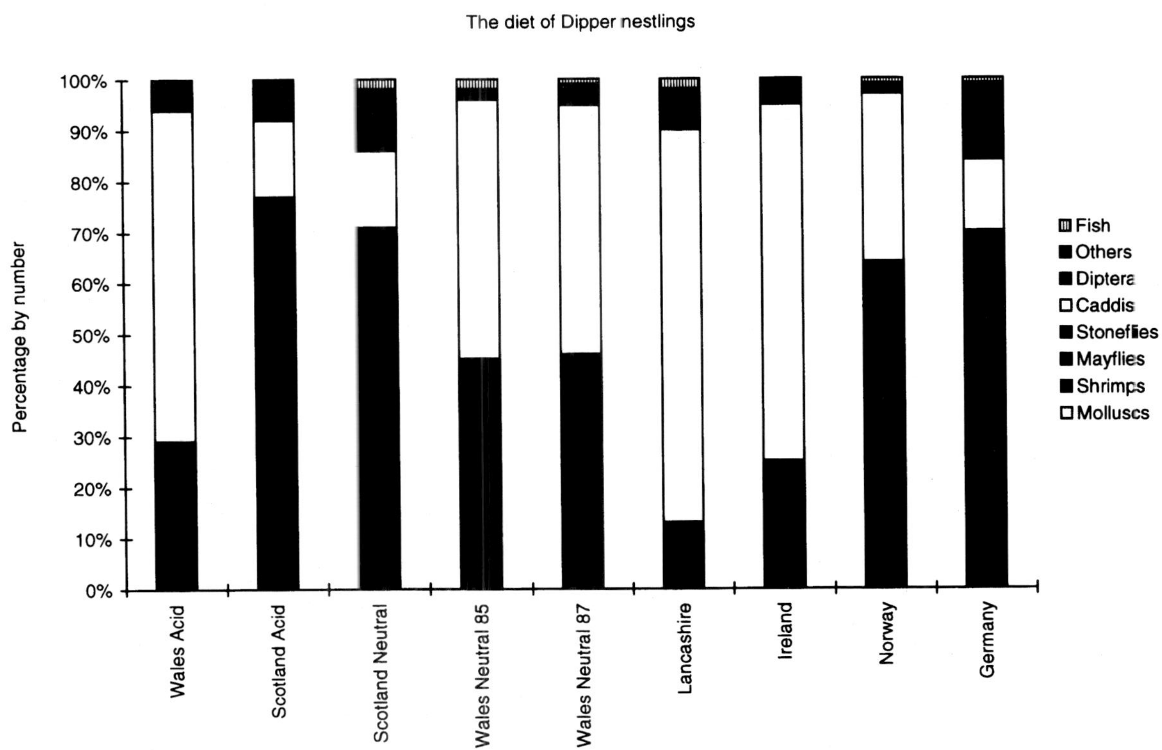

Figure 23. *(a) The diet of Dippers in the spring (breeding season). (b) Diet of nestling Dippers.*

NOTES: From studies in Wales (Ormerod, 1985a), Scotland (Vickery, 1988), Norway (Ormerod, Efleland and Gabrielsen, 1987), Lancashire (Ormerod, 1985b), Spain (Santamarina, 1990), Ireland (Ormerod and Perry, 1985), Germany (Jost, 1975a) and Hungary (Vollnhofer, 1906).

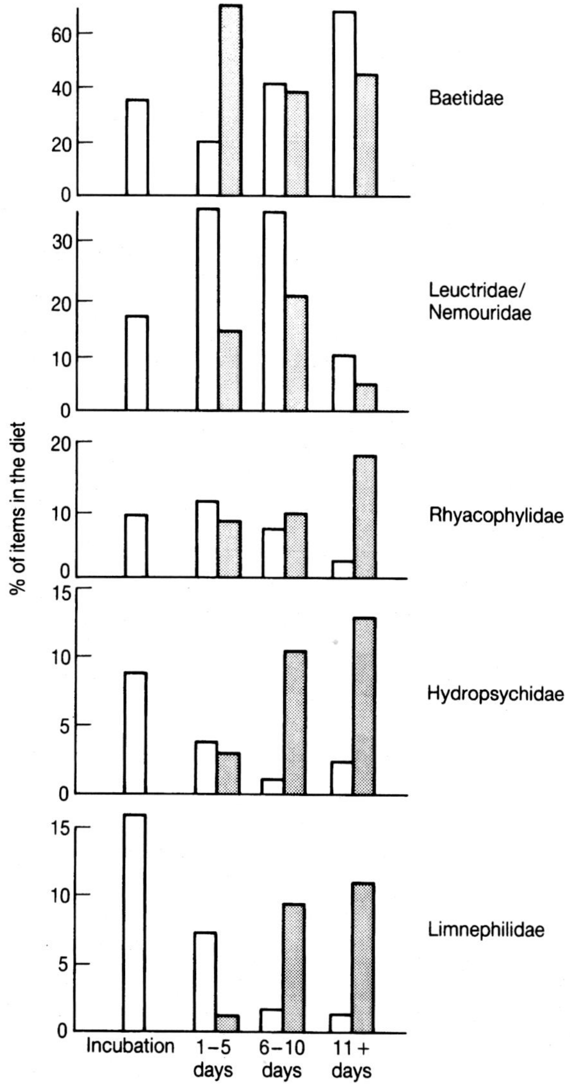

Figure 24. Changes with nestling age (days) in the percentage contribution by invertebrates to the diets of adult (clear) and nestling Dippers (shaded). Source: Ormerod, Efteland and Gabrielsen (1987).

some sites the larvae of moths such as *Tortrix virida* fall into river pools overhung by oaks and other broadleaved trees. Dippers in several locations use these prey in feeding nestlings, clearly no coincidence in view of their morpho-

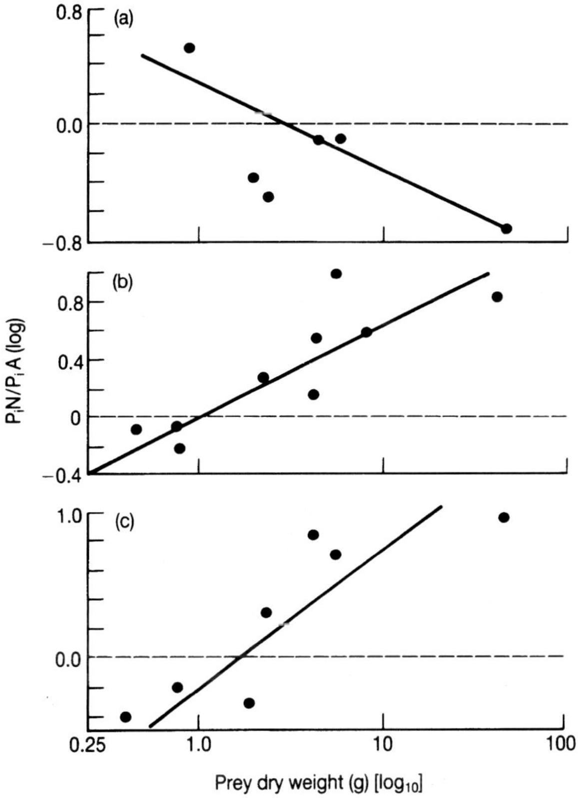

Figure 25. Relationships between the ratio P_iN/P_iA *and prey weight for three nestling age intervals.*

NOTES: (a) 1–5 days old ($r_s = -0.521$, NS); (b) 6–10 days old ($r_s = 0.881$, $P < 0.01$); (c) 11+ days old ($r_s = 0.928$, $P < 0.001$). P_iN and P_iA are the percentage of prey type i in the diet of nestlings and adults, respectively. Source: Ormerod, Efteland and Gabrielsen (1987).

logical similarity to caddis larvae. In these instances the Dippers are freed, too, from the burden of extracting the larvae from cases.

DIET DURING MOULT (see also Chapter 7)

During the moult, some birds require a special diet to help them provide the particular proteins needed for feather formation. This does not appear to be the case in Dippers, however. The diet at this time is rather similar to that in the winter with caddis, mayflies and blackflies dominating. Stoneflies are again largely avoided (Fig. 27).

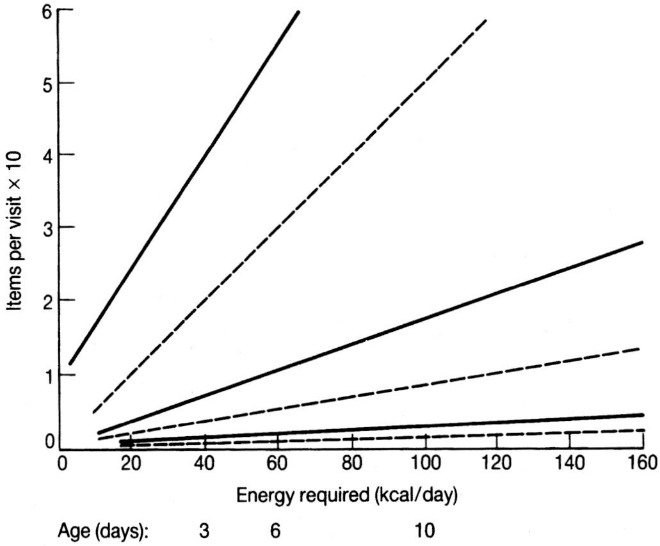

Figure 26. Theoretical values for the number of items required per visit to satisfy given energy requirements of a brood of five Dippers.

NOTES: Alternative models have been suggested for two-visit regimes (broken line = 400 visits per day, solid line = 200 visits per day) and for three prey sizes (1, 6 and 30 mg dry weight). Likely energy requirements were calculated from Kendeigh, Dol'nik and Gavrilov (1977) using brood weights from Ormerod *et al.* (1986). Invertebrate dry weights were converted to energy using a value of 4.8 kcal/g given by Cummins and Wuycheck (1971) for aquatic insects. (After Ormerod, Efteland and Gabrielsen, 1987.)

Table 5. The mean individual weight of prey (within families) eaten by breeding adult Dippers and nestlings over 5 days

	Norway[1]			*Scotland*[2]		
Family	*Adults*	*Nestlings*	P	*Adults*	*Nestlings*	P
Baetidae	1.1	1.3	NS	1.2	1.8	0.001
Leuctridae/Nemouridae	2.2	2.2	NS	1.0	2.7	0.01
Perlidae/Perlodidae	1.2	2.5	0.001	2.4	1.5	0.01
Rhyacophilidae	4.2	4.7	NS	6.4	7.6	0.05
Hydropsychidae	5.7	6.8	NS	4.3	4.1	NS
Limnephilidae	47.4	38.5	0.05	9.3	24.3	

NOTES:
The weights were reconstructed from mouthpart sizes, and the probability of a significant difference is shown according to *t* tests.
[1]Ormerod, Efteland and Gabrielsen, 1987. [2]Vickery, 1988.

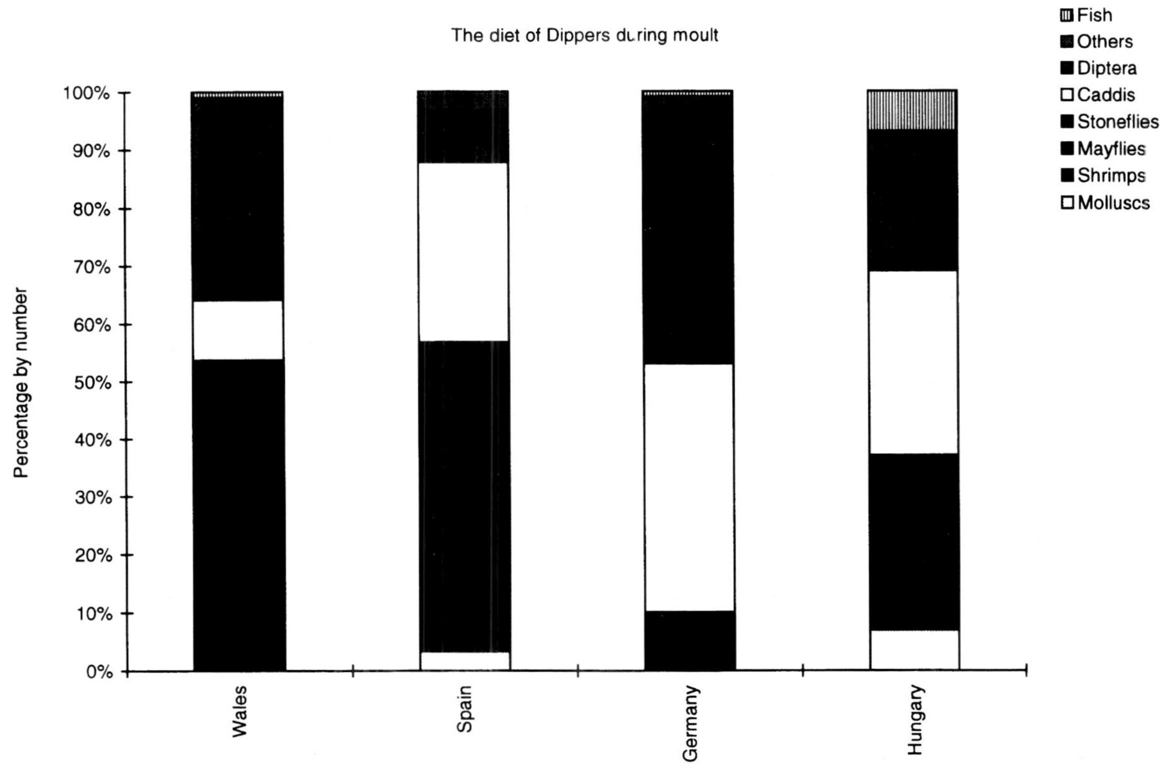

Figure 27. The diet of Dippers during moult. (Sources as in Figs 23 and 24.)

Winter Diet

It is in the winter that the diet varies most from site to site and region to region (Fig. 28). At this time, the birds show less preference than at other times of the year, feeding opportunistically on whatever is available. Nevertheless, large prey such as caddis larvae may still be represented in the diet to a greater extent than would be expected from their abundance in the stream (Spitznagel, 1985c). Caddis are a staple component of the diet of Dippers everywhere, especially

Simuliid *(blackfly) larvae*

when assessed by weight (Table 4). Studies in Wales, Ireland, Avon, Morocco and Spain (Santamarina, 1990; Ormerod and Tyler, 1991a; Tyler and Ormerod, 1991) show that blackfly larvae, mayflies and caddis are the most common items, with stoneflies eaten only rarely. Those stonefly nymphs that are taken are often large carnivorous genera which sometimes hunt over the river bed. They are thus more accessible to Dippers than other members of this group which are more usually hidden in the interstices between stones, or in leaf piles, and hence may not often be encountered by foraging Dippers (Fig. 29). This is an interesting example of how the microdistribution of a prey item may influence its selection. On acidic streams other prey are scarce and stoneflies are represented more in the Dipper's diet, presumably at some cost given the negative selection elsewhere.

The freshwater shrimp *Gammarus* is apparently important to Dippers during the winter in parts of central Europe (e.g. Horvath and Andrikovics, 1991). However, this may in part be a reflection of methodological differences since both Jost (1975a, b) and Spitznagel (1985c) used regurgitated pellets to produce their German data. Nevertheless, Vollnhofer's (1906) Hungarian study was based on gut contents and still showed *Gammarus* dominant. Regurgitated pellets and faecal pellets in Wales gave generally similar patterns of diet, with neither dominated by *Gammarus*.

Fish make their greatest dietary contribution during the winter, then providing up to 5 or 6% of the items eaten; a higher value (7.5%) for a Hungarian

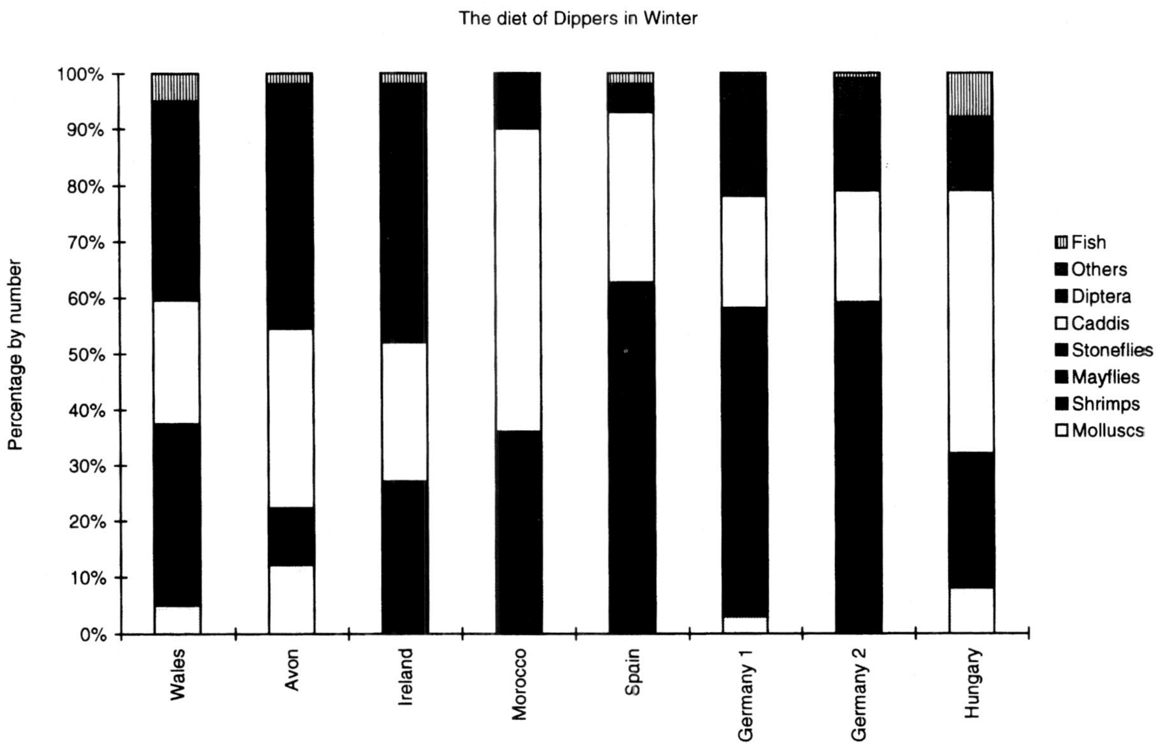

Figure 28. The diet of Dippers in the winter.

NOTES: From studies in Germany (Jost, 1975a; Spitznagel, 1985), Morocco (Tyler and Ormerod, 1991), Wales (Ormerod and Tyler, 1986), Avon (N. White, pers. comm.), Wicklow, Ireland (B. O'Hanrahan, pers. comm.), Spain (Santamarina, 1990) and Hungary (Vollnhofer, 1906).

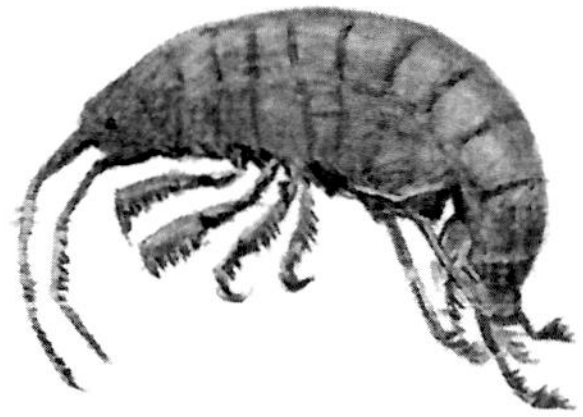

Freshwater shrimp (Gammarus pulex)

Bullhead or miller's thumb (Cottus gobio)

sample was based on the analysis of gut contents which were surprisingly deficient in small prey such as mayflies and blackfly larvae, which elsewhere are very common dietary components. In Wales, bullheads are the commonest fish species taken, comprising up to 80% of the fish eaten. Their small size and benthic habit makes them easy prey for Dippers, and because their density in Welsh rivers is greatest from the autumn onwards following spawning from May to July, they are abundantly available as prey in the winter months. Elsewhere there are many records of Dippers feeding sometimes on the fry or eggs of salmonids (see Chapter 10), but all the quantitative studies indicate that these fish are a minor numerical component of the diet. Nevertheless, fish may be important to Dippers, and during winter in Wales they provide 65% of the total weight of food consumed (Table 4). The use of fish and other calcium-rich prey such as *Gammarus* and molluscs, at this time may also have some importance subsequently in providing a reserve for egg formation (see Chapters 5 and 6).

Partly because of opportunistic foraging during the winter, novel food sources sometimes appear. Vader (1971) observed Dippers foraging during October below the high-tide line in western Norwegian fjords. Here, streams flowing through the brown sea wracks *Fucus* and *Ascophyllum* contained abundant periwinkles *Littorina saxatilis*, isopods *Jaera albifrons* and the shrimps *Gammarus duebeni* and *G. raddachi*, all tolerant of brackish conditions. These were likely to be the major prey items for Dippers here, with interesting ramifications for salt and water regulation. On the Isle of Rhum in Scotland, Dipper faecal pellets collected for us below the high-tide line by the late Wilf Nelson contained

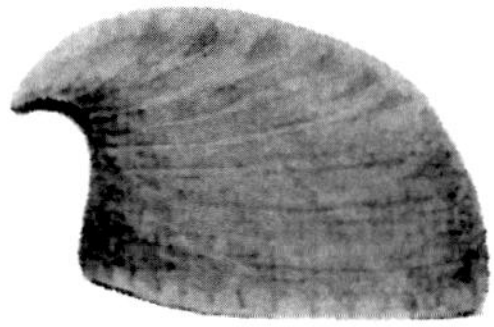

Freshwater limpet (Ancylus fluviatilis)

mostly *Gammarus* and *Orchestia* amphipods, though freshwater blackfly larvae, caddis and stoneflies were still common, indicating that the birds may have been alternating between brackish and freshwater feeding.

Handling Prey

The need for handling different prey varies with their size and characteristics. Small items such as *Baetis* mayflies or blackfly larvae require almost no handling time other than a peck, and concentrations of blackfly larvae are sometimes harvested by a sideways scraping or scything action of the bill. Sonya Yoerg estimated capture and ingestion rates for blackfly of up to three per second, even by juveniles. By contrast, cased caddis larvae such as *Stenophylax* limnephilids are treated by a prolonged process of extraction which appears to follow a predictable course (Jost, 1972). Once the cased caddis is caught, it is taken to dry land, often a mid-river rock, where alternately the case is squeezed with the bill and knocked end-on against a stone several times. When the larva eventually

Dipper extracting a caddis larva from case

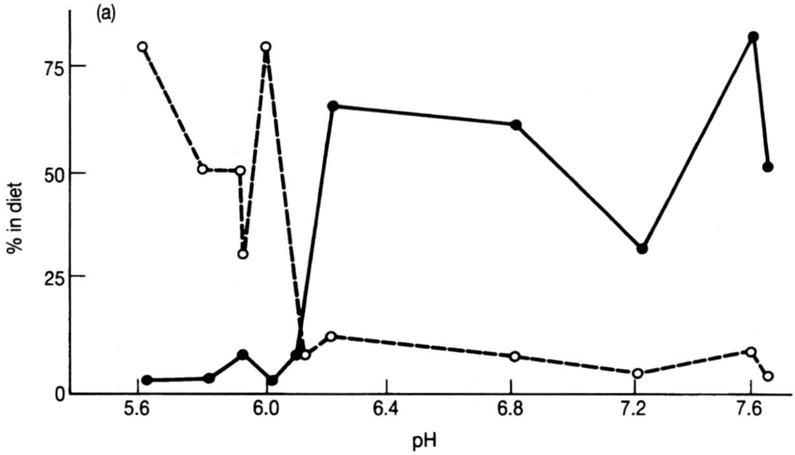
(a)
75
50
25
0
% in diet
5.6
6.0
6.4
6.8
7.2
7.6
pH

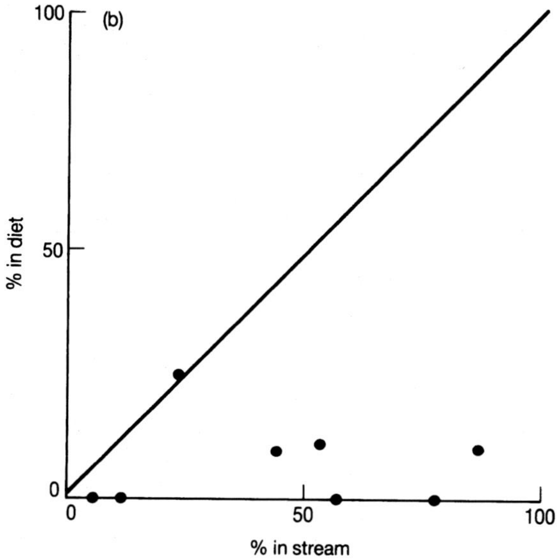
(b)
100
50
0
% in diet
0
50
100
% in stream

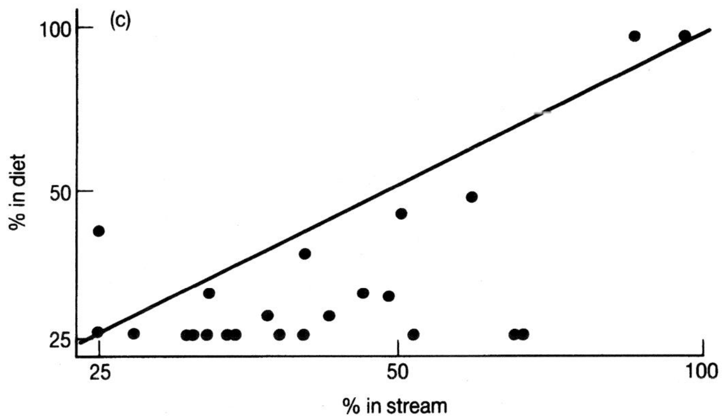

Figure 29. Three different instances of selection against stoneflies, including a case of prey switching.

NOTES: Shown by the percentage contribution made to the diet of: (a) adult and nestling Welsh Dippers by stoneflies (open circle) and mayflies (closed circle) in relation to stream pH (Ormerod and Tyler, 1991); (b) adult Welsh Dippers during moult by stoneflies (Smith and Ormerod, 1986) and (c) Welsh Dippers during the winter by stoneflies (Ormerod and Tyler, 1986).

begins to emerge, it is seized by the Dipper and the case knocked or shaken off. (American and Brown Dippers extract caddis larvae from their cases in an identical way to that shown by Dippers in Wales and elsewhere.) During brood provisioning, when loads of several caddis are often fed to nestlings at each visit, the handling involved in collecting and extracting will demand great dexterity. Remarkably, Sonya Yoerg (pers. comm.) has observed adult Dippers foraging for young perform this capture and extraction three or more times in 10 seconds, although more realistic foraging rates are around two captures of medium to large prey per minute. Fish, too, are sometimes repeatedly beaten before being orientated for swallowing head-first. Detection, capture and handling times for these can allow ingestion of around one per minute, though this may be exceptional.

Food-washing as recorded in the American Dipper by Evenden (1943) is not normal behaviour in Dippers in Wales.

Learning to Forage

The exact degree to which dippers use innate foraging skills, rather than those learned during their lifetimes to suit their particular riverine setting, is difficult to know. Some groups of insects have highly predictable patterns of microdistribution which were probably established long before dippers radiated into the same river environment. In other instances environmental variability will result in differences in prey distribution from territory to territory, and even from time to time. Juvenile dippers almost certainly learn their skills to some extent. Sonya Yoerg, whilst assessing learning behaviour in recently fledged Dippers in mid-Wales (see Chapter 7), recorded over 230 hours and 40,000 prey captures by 13 juveniles and 9 adults. Juveniles less than five days after fledging rested nearly 90% of the time and foraged less than 4%. Over the next four days, a marked transition occurred as they increased their foraging time to 30% (resting 60%), and to 50% by Day 21. This is roughly equivalent to the time spent foraging by adults. Provisioning of food by the adult Dippers to their young declines over the same period, ceasing almost entirely by Day 14. Interestingly, begging behaviour by juveniles occurs mostly over the sixth to ninth day after fledging, although it continues even after their parents have ceased feeding them.

Yoerg noted clear trends in juvenile foraging success. For all but the smallest, easily collected prey (blackfly and *Baetis* mayflies), prey captures per unit time foraging were very low for the first week, but then markedly increased, although juveniles were still less efficient than adults by the time they dispersed (see Chapter 7). For example, a juvenile at two weeks old let about two-thirds of large prey captured fall into the water. There were many instances of juvenile handling cased caddis for 20 seconds or more only to discover them empty! At other times, failure to extract the caddis led to it being eaten together with its case after considerable effort was expended. Even at three weeks of age, juveniles captured only 0.3 medium-sized prey per minute while adults captured two. Capture rates of small prey did not increase, however, suggesting that they were very easy to catch or had a predictable distribution which need not be learned. Juveniles were more successful at foraging in slow-moving water at the river margins than in turbulent water.

FORAGING AND FOOD OF OTHER SPECIES OF DIPPER

Available reports and our own observations of American and Brown Dippers indicate that they both use very similar foraging techniques to the Dipper in Europe and, as noted above, they use similar habitats within the river corridor. However, whilst these three species of dipper frequently swim and dive, the two South American species rarely do so. From limited observations the main foraging techniques of both the White-capped and Rufous-throated Dipper appear to involve wading in river margins, probing among rocks and stones and picking prey from rock surfaces, and standing in the fast current on the lips of cascades and waterfalls and picking prey from these locations (Tyler, 1994).

Our information on the diets of other species of dipper is still scant and confined largely to the American Dipper. However, since hill streams have very similar faunas across most of the world, and because the adaptations of the

Table 6. The diets of American Dippers in different locations and at different times of the year

	Breeding			*Winter*
	Alberta[1] (n = 6)	*W Montana*[2] (n = 26)		*SW Washington*[3] (n = 6)
		River 1	*River 2*	
Molluscs	0	0	0	4.4
Mayflies	66.3	13.5	51.7	0.9
Stoneflies	5.7	9.1	1.2	3.9
Caddis larvae	27.8	76.8	23.6	43.2
Coleoptera	0	0.3	0.2	3.7
Diptera	0	0.3	23.3	40.2
Fish	0	Present	0	0
Total items	276	1008	606	876

NOTES:
The values are percentage contributions by number to the total animals in the guts of all birds combined.
[1]Ealey, 1977. [2]Mitchell, 1968. [3]Thut, 1970.

different dipper species are nearly identical (save for the lack of diving in the two South American species) major dietary differences are unlikely. This view is supported by the data we have for the American Dipper, of which 38 birds were shot in four different locations for the quantitative analysis of diet. This showed that mayflies and caddis dominated the diet during breeding, with caddis, mayflies and dipteran larvae (mostly chironomids) eaten during the winter (Table 6). Caddis were selected in proportions greater than expected from the abundances in the streams while mayflies, and particularly stoneflies, were selected against (though there were differences within these groups). The occurrence of all these groups in the diet provides a strong consistency with Europe.

Our limited observations of Brown (Tyler and Ormerod, 1994), White-capped and Rufous-throated Dippers indicate that they too prey largely on caddis larvae, mayfly nymphs and blackfly larvae. However, Sergio Lasso (pers. comm.) sometimes observed White-capped Dippers foraging from vegetation on river banks, apparently taking terrestrial prey such as earthworms.

ECOLOGICAL IMPACTS

As a result of their energy demands and their predatory load on fixed territories, some questions have arisen over the extent to which dippers influence the abundances of their prey. In Wales, we attempted to answer this question by assessing energy requirements through the year, and placing them in the context of what we knew about Dipper diet. Using representative territory sizes of 4680–11,250 m^2, we estimated that a pair of Dippers resident all year and rearing one or two broods, could exploit about 1.0–2.4 g dry weight per m^2. However, because this predatory load is focussed on certain prey such as caddis larvae and fish, predatory losses in these groups could be 0.5–1.1 g per m^2 and 0.8 g per m^2, respectively. Unfortunately we do not have contemporaneous data on production against which to judge these losses, although available data indicate that bullhead production in Welsh streams is up to 0.5 g dry weight per m^2. Predation by Dippers could thus account for a substantial amount of this, with potential impacts on bullhead populations (Ormerod and Tyler, 1991a). By similar reasoning, Dippers could account for a substantial amount of the production of limnephilid and hydropsychid caddis. Together, these calculations indicate at least one of the reasons why Dippers need to defend territories of 4000 m^2 or more.

Dippers roosting under bridge

CHAPTER 5

Winter and early spring behaviour

It is difficult to define the winter for resident Dippers because their breeding season begins so early in the year, but in this chapter we look at different wintering strategies and the extent to which Dippers establish winter territories. We then discuss roosting behaviour, song and courtship, different mating systems and nest-building. First it is relevant to consider size differences between the sexes and sex ratios of dippers because these are factors that will affect territoriality and mating systems, as well as survival.

SEXUAL DIMORPHISM AND SEX RATIOS

White-throated Dippers are sexually dimorphic, with males being larger than females (Andersson and Wester, 1971; Svensson, 1992). The size difference is not always apparent in the field, but in the hand wing measurements and weights readily separate the majority of birds. Males are heavier (Fig. 30) and have

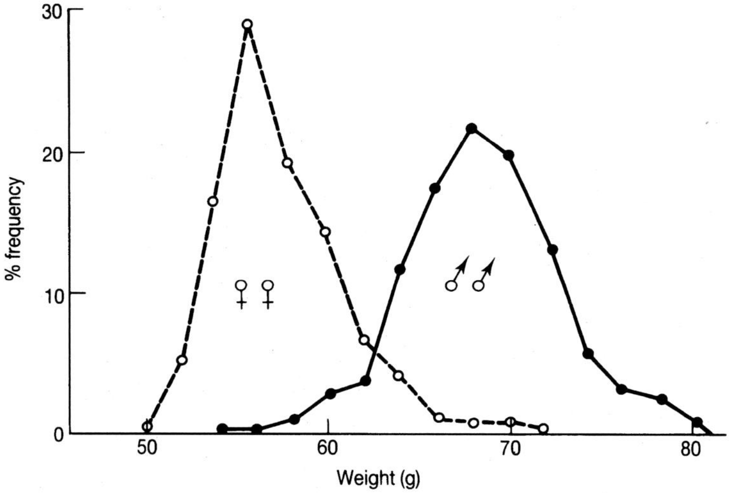

Figure 30. Percentage frequency polygons showing the weight distributions of 259 male (●) and 259 female (○) Dippers in Wales.

NOTES: Mean weight of male Dippers = 67.90 ± 0.48 g, 95% cI; mean weight of female Dippers = 57.26 ± 0.49 g, 95% cI. Dippers were sexed according to wing length (Ormerod, Tyler and Lewis, 1986).

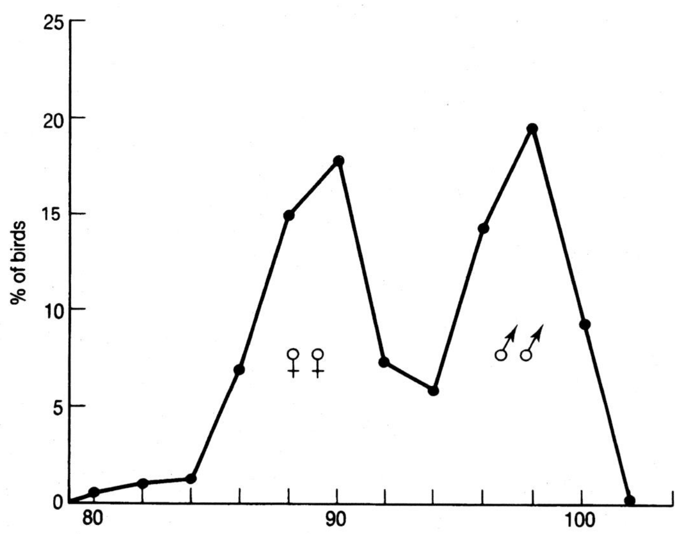

Figure 31. Percentage frequency polygon showing the distribution of Dipper wing lengths in Wales (Ormerod, Tyler and Lewis 1986).

significantly longer wings (Fig. 31) and tarsi than females although there is some overlap between small males and large females. For our Welsh population this overlap in wing length is between 92 and 94 mm with males having a mean wing length of 96.7 mm and females 88.3 mm (Ormerod, Tyler and Lewis, 1986). In Ireland, O'Halloran, Smiddy and O'Mahony (1993) found that the degree of sexual dimorphism was less than expected, females being larger than reported from most other studies. In Wales and Ireland first-year birds, both males and females have slightly shorter wings than older birds (Fig. 32). In Germany, however, Schmid and Spitznagel (1985) found that although first-year males had shorter wings than older birds, wings of females were not significantly different in length in young and old birds.

Both Irish and Welsh male Dippers have slightly longer wings than birds measured in Scotland by Galbraith and Broadley (1980) (Table 7), contrary to what one might expect from Bergmann's rule (body size increasing at higher latitudes). Birds of the nominate race *cinclus* in Scandinavia do, however, have longer wings than *gularis* in Wales or Scotland (Andersson and Wester, 1971; Lundberg, Bergman and Olsson, 1981). The mean wing length of male Dippers in northern Sweden (98.6 mm) is also significantly greater than that of male Dippers (*hibernicus*) in south-west Ireland (97.8 mm), but there is no significant difference in mean wing lengths of female Irish (89.8 mm) and Swedish

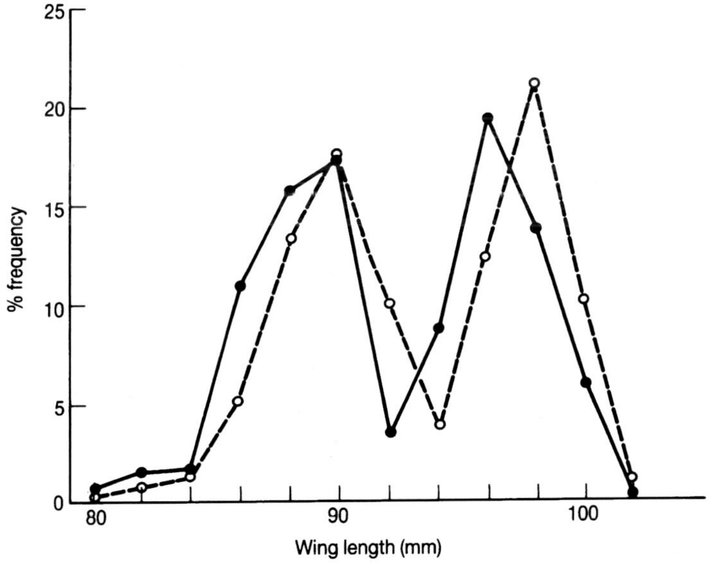

Figure 32. Percentage frequency polygons showing the distribution of wing lengths of first-year Dippers (●; n = 224) and Dippers over one year old in Wales (○; n = 320) (Ormerod, Tyler and Lewis 1986).

Table 7. Differences between races of C. cinclus *in mean wing length of males and females*

Race	Country	Mean wing length (mm) Males	Females	Source
gularis	Wales	96.7 (282)	88.31 (270)	Ormerod *et al.* (1986)
gularis	Scotland	96.0 (89)	88.1 (81)	Galbraith and Broadley (1980)
hibernicus	SW Ireland	97.8 (280)	89.8 (246)	O'Halloran, Smiddy and O'Mahoney (1992)
cinclus	N Sweden	98.6	90.0	Lundberg, Bergman and Olsson (1981)
cinclus	S Sweden	98.41 (266)	89.79 (358)	Andersson and Wester (1971)
aquaticus	Germany	95.07 (173)	87.01 (152)	Rockenbauch (1985b)
aquaticus	Germany	94.7 (91)	86.83 (69)	Schmid and Spitznagel (1985)
minor	Morocco	102.8 (9)	94.3 (3)	Tyler and Ormerod (1991)

Ranges for wing lengths of males and females of *leucogaster* and eastern races are given by Ali and Ripley (1983).

(90.0 mm) Dippers (O'Halloran, Smiddy and O'Mahony, 1993). The relatively large size of Irish and Welsh birds may be due to genetic isolation of these populations, or to other factors such as the high quality of territories in much of our Welsh study area. A difference in the ways birds are measured cannot be ruled out; we use maximum flattened chord (Svensson, 1992) but in some studies it is unclear whether this method was used. Moreover, wing lengths of dry museum specimens are likely to be shorter than those of live birds.

A geographical variation in body size which occurs in four different populations in south-western Germany (Schmid and Spitznagel, 1985) corresponds to Bergmann's rule. Here the difference in size between the sexes was also greater towards the north, and Schmid and Spitznagel suggested that this was an adaptation to the more severe conditions at higher latitudes. It was, they believed, advantageous for females to be smaller, requiring less food, whereas males through sexual selection became larger. Other races vary in size but large birds as occur in Morocco and in Iran (see section on races in Chapter 1, pp. 8–12) might seem to contradict any pattern of size increasing to the north. However, high altitudes may also result in larger body size.

In Wales, we found a marked seasonal variation in weights with those of males increasing between October and February with peaks just prior to breeding (Fig. 33). Male weights declined during the breeding season, increasing again during moult. Likewise females increased in weight during the winter up to egg-laying, prior to which they were particularly heavy, and remained heavy in March and April (Fig. 34). Their weights then declined until July but, as with males, increased again during moult. First-year birds of both sexes declined in weight in the autumn, but gained weight during the winter (Fig. 34). Galbraith and Broadley (1980) described a similar pattern in seasonal weight changes in

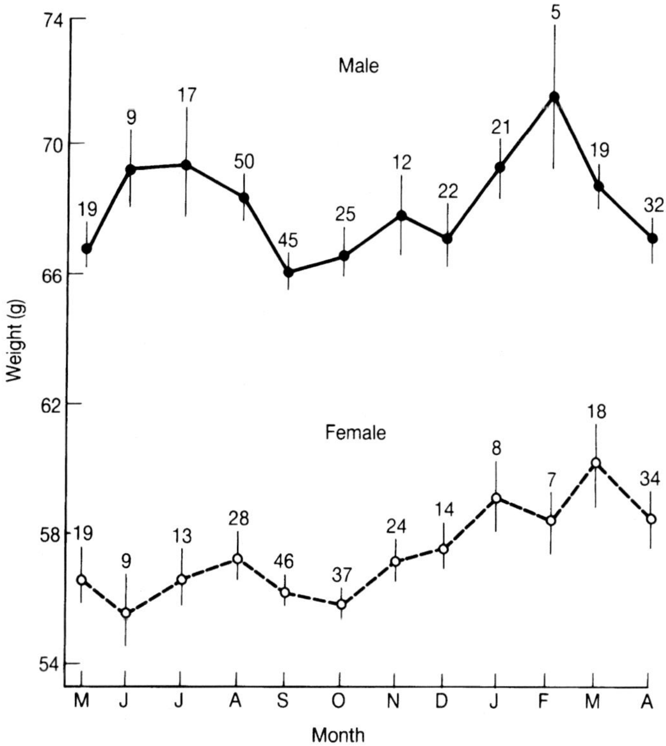

Figure 33. Annual weight changes (mean ± SE, headed by sample size) of male (top) and female (bottom) Dippers in mid- and south-east Wales.

Dippers in Scotland to that in Wales. By contrast, seasonal changes were found to be less marked in Irish Dippers, although in females weight did peak during the egg-laying period and decreased after the breeding season reaching its lowest in September (O'Halloran, Smiddy and O'Mahony, 1993). The reasons for these differences are not clear.

These data show that despite some adverse weather, Dippers in Britain and Ireland are able to gain weight during the winter. By February and March they are in good condition for territory formation, courtship, nest-building, egg-laying and the rearing of nestlings. By contrast, migratory Dippers of the nominate race only gain weight at the end of February and in March (Andersson and Wester 1972; Lehikoinen and Hakala, 1988), with females and first-year birds showing little weight increase before April (Lundberg, Bergman and Olsson, 1981). The differences in timing of the breeding between Scandinavian and British birds might in part be a consequence of such differences in body condition.

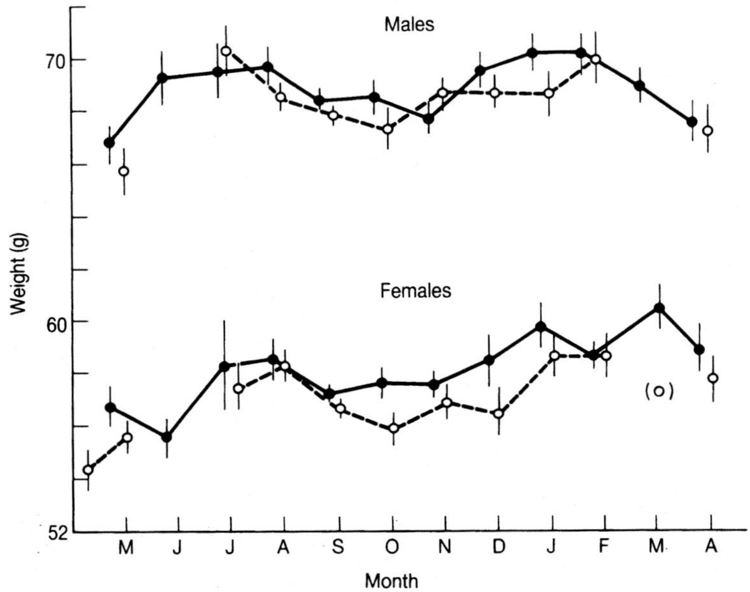

Figure 34. Annual weight changes in first year (○) and adult (●) male (top) and female (bottom) Dippers in our Welsh study area.

In a large sample of Dippers in our study area and in Ireland where birds are largely sedentary, the sex ratio was close to unity (Ormerod, Tyler and Lewis, 1986; O'Halloran, Smiddy and O'Mahony, 1993). Other studies have shown both males (Richter, 1953; Preuss, 1959; Lundberg, Bergman and Olsson, 1981) and females to be in excess (Andersson and Wester, 1973; Lagerstrom, 1979). These have, however, involved either small samples or were undertaken in winter when sex-specific patterns of migration produce heavy bias. In very severe winters or at other times of stress, it is likely that the larger males suffer heavier mortality, leading at times to females outnumbering males during the breeding season. On some rivers and in some years in our study area this has been the case.

Sexual Dimorphism in Other Species of Dipper

The degree to which the other four species are sexually dimorphic in size is unclear. Ali and Ripley (1983) record male Brown Dippers as being larger than females. The very small samples that we have caught of Brown and Rufous-throated Dippers (Tyler, 1994; Tyler and Ormerod, 1994) also show that females are smaller than males but that this difference is less marked than in the Eurasian species.

WINTER TERRITORIALITY

There are conflicting reports about the territoriality of Dippers during the winter. Creutz (1966) reported individuals and pairs showing territorial behaviour at this time, and Preuss (1959) found that migrant Dippers wintering in Denmark maintained strongly delineated territories, except just after arrival and again before departure. He observed many territorial fights, with both birds stretching necks, one bending legs and flapping wings and then flying at the other, as well as attacks from the air. Likewise Holmbring and Kjedemar (1968) noted that Dippers in the province of Ostergotland in Sweden maintained winter territories.

By contrast, Lundberg, Bergman and Olsson (1981) found no cases of true territories in a wintering population in coastal northern Sweden, but they did record fights and aggressive behaviour between birds and suggested that this was in order to maintain the dominance system in each stream population.

Cousins (1985) observed different degrees of territoriality in a wintering population on the River Devon in Clackmannanshire in Scotland, ranging from exclusive site defence, to defence of an individual space to non-territorial behaviour. In general males were most aggressive and dominant, and exclusively territorial birds were usually males. Uses of a model Dipper caused males to behave more aggressively whereas a tape of Dipper song played with the model caused females to show greater avoidance of apparently unoccupied areas. Non-territorial birds were almost always females.

Wintering strategies depend very much on geographical locality and the severity of the weather. As we discuss in Chapter 7, the most northerly Dippers are migrants, flying up to 1000 km south for the winter; some are altitudinal migrants whilst others, in milder areas, remain on or close to their breeding grounds, moving away only in very harsh weather. Those that remain on their breeding rivers where open water and a plentiful food supply are assured throughout the winter, establish territories during the winter either as pairs or as single birds; these winter territories may be the same as their breeding territories. Where many birds move down on to lower streams or where partial freezing over is likely, concentrations will inevitably occur. Then in extreme cases where birds are closely packed, only an individual space may be defended, or alternatively, territories will be small. The key to whether birds show territorial defence or defence of only a small area around themselves would therefore seem to depend partly on the sex of the bird (Cousins, 1985) but largely on the nature of the stream and the severity of winter weather in the area.

In Britain many observers have reported Dippers showing territorial behaviour in the winter. Murphy (1972) found winter territories to be established in September to October and to exist until February or even March. In much of our study area, winter territories are very similar to breeding territories, the two often coinciding, with many pairs remaining on the same stretch of 500–1000 m of river for the whole of the year. There are some single birds during the winter, some of these occurring on rivers where breeding does not occur. On poor quality, acidic streams winter abundances are low, *c.* 0.25 birds/km; elsewhere in Wales they may commonly be 2–3 birds/km. On the R. Wharfe in Yorkshire Marjorie Andrews (pers. comm.) recorded up to 2.3 birds/km during October and November.

In Germany Richter (1953) found that birds wintered either in pairs or as solitary male and female. Some remained on their breeding territories, or moved only a short distance downstream, often returning to their breeding territories at dusk to roost; others moved completely away.

In parts of Germany winter abundances of Dippers are comparable to those in Wales, i.e. 1–2 birds/km (Moll, 1977; Wink and Gerstberger, 1977; Schneider, 1985). Schneider found altitude had little effect on wintering abundances, birds occurring from 150 to 1000 m, and 68.5% of his birds were paired during the winter. The mean territory length of single birds was shown by Holmbring and Kjedemar (1968) to be smaller (350 m) than that of pairs (440 m). In southern Sweden, Fredriksson and Gotmark (1982) found that Dippers of the nominate race normally hold territories but when the upper parts of streams freeze over, birds move lower, extending their home range to over 3 km. Many birds disappear, either dying or migrating further south.

In areas where streams regularly freeze over, Dippers move downriver in the autumn, sometimes well away from their breeding territories. Severe spells of weather can cause temporary concentrations of birds within short stretches of unfrozen water, resulting in a breakdown of territorial behaviour and defence by Dippers of only 'individual distances'. Thus Alder (1966) recorded two birds at the same ice-hole feeding together, whilst Murray (1967) recorded three Dippers together, diving into a patch of open water at an otherwise ice-covered loch in Perthshire, Scotland. On the R. Spey in Scotland, 13 Dippers were recorded by Perry (1976) on a 500-m bend in the river in mid-November. The number had increased to 16 by early December. Some 20 birds, including this concentration on the bend, were then counted on just 1 km of river. Frequent pursuit flights and threat displays were observed, but Perry concluded that each bird defended one or more small patches at the water's edge. When the river partly froze in early January, birds concentrated by remaining open water, with 8 observed lined up on either side of a narrow *c.* 30-m long channel through the ice. Here any two neighbours often made threat displays, one bowing and wing-flicking and the other in the upstretched posture, and indulged in pursuit flights, each defending a space along the edge of the ice. Watson (1972) quotes a report of as many as 100 Dippers on 1.6 km of the River Ewe in Scotland on 31 January 1879.

Similar aggregations of continental Dippers have been recorded. For example, in the Harz Mountains of western Germany, Haensel (1977) observed large concentrations of up to 26 birds, whilst in the Lübeck area of Germany 7 birds were seen on a 300-m stretch of river, feeding amicably, when concentrated by harsh conditions elsewhere (Arndt, 1979).

Where conditions allow, some birds clearly do maintain winter territories even where concentrated on these lower streams, but these territories may only be 100–250 m in length (Vogt, 1944; Schuster, 1953; Balat, 1962b; Sudhaus, 1972; Kaiser, 1988).

American Dippers, according to many observers (e.g. Bakus, 1959b), maintain winter territories but shifts downriver may occur in severe weather. Price and Bock (1983), however, found little evidence of winter territoriality amongst American Dippers in their Colorado study area. Most birds were mobile and left the area for at least part of the winter. They recorded four different patterns of movements: some dippers were largely resident, moving as little as necessary to find open water; several apparently made regular trips to wintering areas; a few stayed on the study area but wandered at random; whilst the fourth group were absent for up to five months during the autumn and winter.

ROOSTING BEHAVIOUR

Roosting behaviour is considered here because although Dippers may be territorial throughout the year and usually feed in a solitary way, several individuals or as many as a dozen sometimes gather at sites to roost in the autumn and winter (e.g. Hewson, 1969). In the breeding season birds roost in or close to their nests; some do this too in the autumn and winter, members of a pair roosting singly within their territory on cliff ledges, in rock crevices or other natural sites, in or on their nests or under a local bridge. Others, however, fly to bridges outside their territories to roost communally. Bridges may offer a multitude of roosting opportunities from crevices in stonework, drainpipes in their walls to a series of ledges, girders or struts. When two or more Dippers roost under the same bridge, they perch separately from each other, on different ledges or in different crevices or pipes. Only very rarely have we found Dippers huddling together, and on these few instances the birds involved have been a pair in the same drainpipe.

In the late 1980s we looked at almost 100 bridge roost sites in Wales over more than 300 night-time visits, mainly during the autumn and winter, to assess the relative frequency of gregarious roosting (Ormerod and Tyler, 1990a). We also caught and marked many individuals so that roost site fidelity and day-time feeding areas could be shown. To catch birds at roost is relatively easy. You can sometimes dazzle a bird with a strong torch beam and then pick it off its perch; another strategy is to first locate a roosting bird by torchlight, switch off the light and approach it quietly and then catch it, perhaps with the help of an assistant shining a light at the last minute. Birds roosting in old nests or pipes are easily trapped compared with those on open ledges. Disturbed birds or birds already caught may call, awaking and alerting others, so making them more difficult to catch, but on dark nights birds are less likely to fly from the roosts. A butterfly net can be useful for catching birds perching high up on a beam or ledge. If a bird does escape from the roost before it can be caught, it generally flies only a short distance or drops down into the water, most very soon returning to their roost site with a semi-hovering flight. After a captured bird has been ringed or been weighed and measured, it is gently replaced on or in its roost site with any torches switched off. Most soon settle although some, particularly in strong moonlight, will fly from the roost. We have ample evidence from recapturing numerous birds in successive years that the brief disturbance that roost ringing involves has no detrimental impact on Dippers.

We found that although up to nine birds were present at some of our Welsh roost sites, the majority of sites held only one bird (Fig. 35a). Most Dippers,

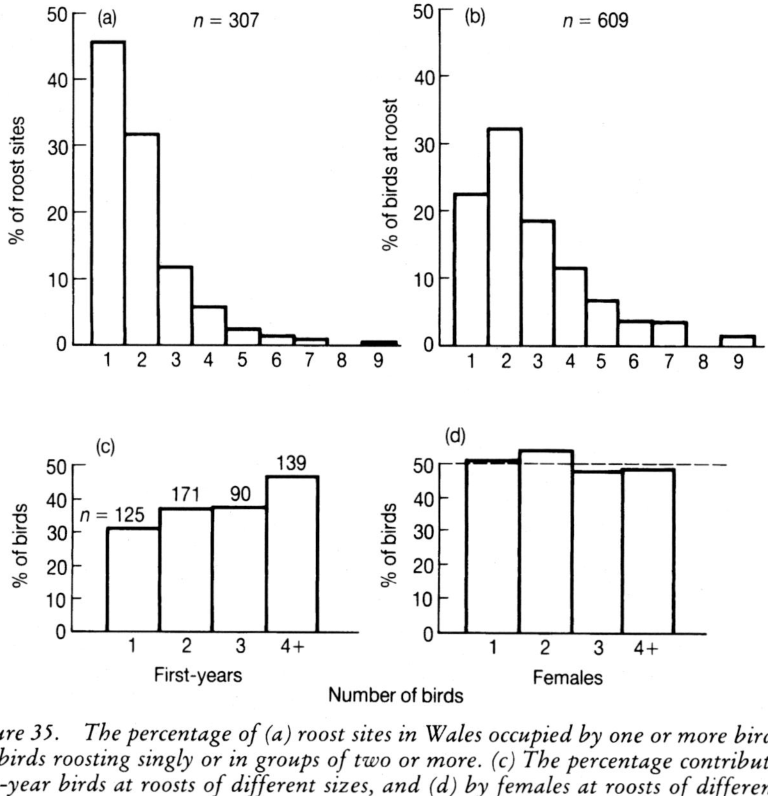

Figure 35. The percentage of (a) roost sites in Wales occupied by one or more birds, and (b) birds roosting singly or in groups of two or more. (c) The percentage contribution by first-year birds at roosts of different sizes, and (d) by females at roosts of different sizes (Ormerod and Tyler, 1990a).

though, roosted in twos (Fig. 35b), the majority of which (65%) were pairs consisting of a male and female. Sex ratios in roosts were around 1:1 in roosts of all sizes (Fig. 35d), although first-year birds made a proportionately greater contribution to larger roosts (Fig. 35c).

Roost size increased during the winter when over 60% of birds were in groups of three or more (Figs. 36 and 37). By February or March, most of our Dippers were again roosting in pairs. There was no evidence that young birds caused the winter increase in roost size because their percentage contribution to groups was lowest at this period. Instead, roost enlargement was probably because at this time birds from natural roost sites moved to bridges, the latter offering warmer conditions than the air temperature in the open during cold periods (Fig. 38). At six bridge sites where we measured temperatures, minimum roost temperatures varied almost directly with air temperatures, but the bridges were warmer than the air on all except three nights, this effect increasing as air temperature fell. Moreover, the bridge with the highest minimum temperature held most birds, whilst the coldest bridge was vacated during the winter. Shaw (1979) also suggested that the microclimate below bridges made them a favoured roost site on cold, windy winter nights.

Although bridges probably do attract more birds in the winter because of their shelter and comparative warmth, the number of birds roosting at each of our sites

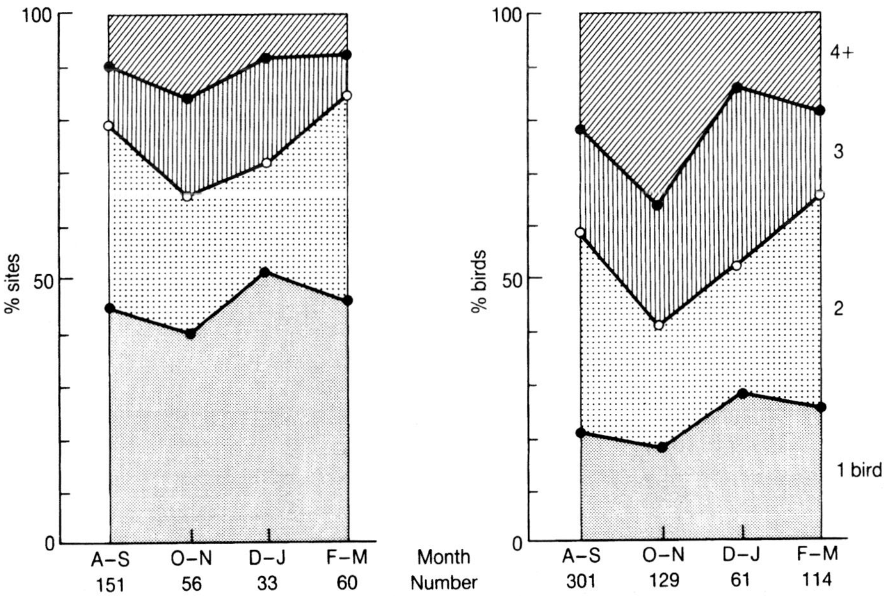

Figure 36. Changes in the percentage of sites and of Dippers in Welsh roosts of different sizes during the autumn and winter. Sample sizes are shown (Ormerod and Tyler, 1990a).

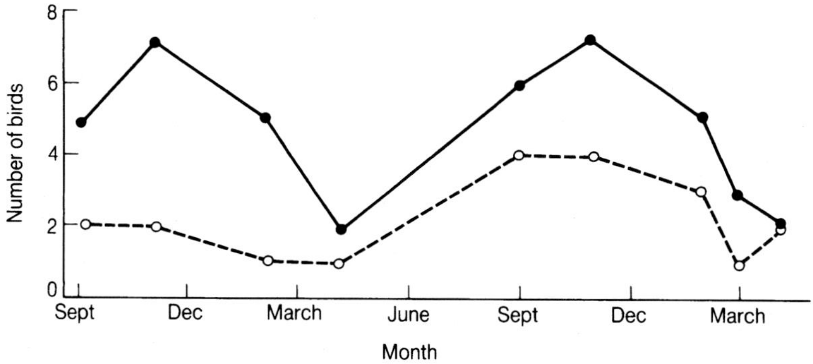

Figure 37. Seasonal changes in the numbers of Dippers roosting at two sites in the catchment of the River Wye (Ormerod and Tyler, 1990a).

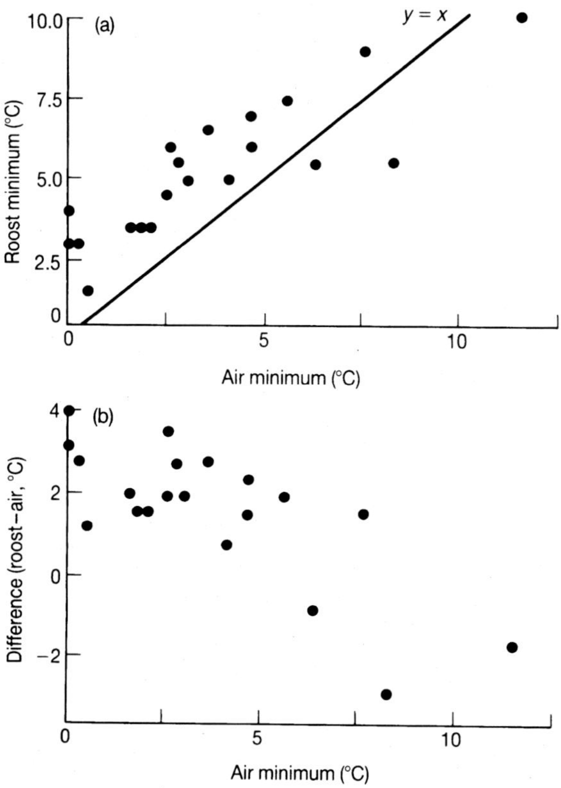

Figure 38. (a) The minimum temperature at a bridge site in Wales used by roosting Dippers in relation to the minimum air temperature at an adjacent site on 20 nights during February and March. (b) The difference between the bridge and air minima in relation to air temperature (Ormerod and Tyler, 1990a).

could partly be explained by the number of niches available (Fig. 39), the local density of Dippers able to fill them, and the distance to alternative locations. On rivers with a wide range of suitable roost sites, any one site rarely held large numbers of Dippers. By contrast, on rivers with few potential roost sites, any bridge with ledges or girders might be used by many birds. For example, on the R. Theme, where few natural roost sites occur, and where there are long stretches without a suitable bridge for roosting, Tony Cross (pers. comm.) caught 19 Dippers in the late summer at just two good bridges: 10 birds were roosting at one of the bridges and 9 at the other, which was only about 1 km from the first.

We found, as did Shaw (1979), that most birds at bridge sites showed high roost site fidelity during the autumn and winter and between seasons (Ormerod and Tyler, 1990a). Continued trapping at roosts since our earlier study has confirmed this. Thus a further analysis of Dippers first caught at roost, between the late summer and mid-winter, and retrapped during the same period in subsequent years up to 1993, showed that about 94% of the birds (n = 195) were faithful to their roost site over periods of up to six years, the maximum life span of the great majority of Dippers (Table 8). Our own longevity record is of a young female first caught at roost in July 1984 in a crevice in a bridge, which then bred and roosted in this site in all seasons for over seven years up to April 1991. We have many other retrap histories for birds caught at the same roost for five or six consecutive years. Only 11 birds (less than 6%) changed roost between years. More than half of these were first caught as juveniles, presumably whilst still dispersing, and had then moved to another roost from 1 to almost 30 km from their original roost, by their second autumn/winter. Several adults did change roost, though, after apparently being faithful to their initial roost for three or more years; these only moved to sites within 5 km of the first roost.

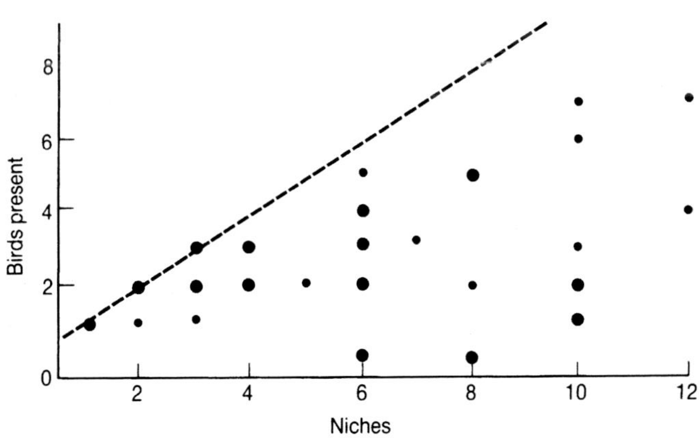

Figure 39. The number of Dippers roosting at 56 bridge sites in Wales in relation to the number of roosting niches available.

NOTES: The line shows $y = x$; small dot = 1 point, larger dot >1 point. See Table 1 in Ormerod and Tyler (1990a) for regression parameters.

Table 8. Fidelity to roost site as shown by the number of Dippers initially caught at roost in Wales and subsequently retrapped at the same or a different roost site

	Retrapped after:					
	1 year	*2 years*	*3 years*	*4 years*	*5 years*	*6 years*
SAME SITE						
Male	33	19	14	8	3	2
Female	55	23	13	5	6	1
Total	88	42	27	13	9	3
DIFFERENT SITE						
Male	3	0	2	0	1	0
Female	4	0	2	0	0	0
Total	7	0	4	0	1	0

Most birds at communal roosts come from the territory containing the roost or from adjacent territories. Almost 95% of 109 birds which had been caught at roost and then recaptured in nets during daylight, or vice versa, were within 3 km of the roost site, with 72.5% being within 1 km. Four other birds were caught up to 5 km away, and a further two up to 8 km away, having overflown territories of neighbouring birds to reach the roost site. The rest had moved further but these were mainly young birds and probably still dispersing when first caught.

Evidence for any communal function, such as an anti-predator value, at gregarious roosts is slight. Although many birds, notably males, did call when caught, other birds at the roost usually took no action. We suggest that the communal roosts are therefore chance aggregations around a relatively scarce resource, namely safe and warm roosting niches.

SONG AND COURTSHIP

By January or February resident Dippers are singing vigorously, courting and building new nests or repairing old ones. Altitudinal migrants gradually move back up to their breeding streams during February and March with true migrants returning north from the end of March.

Song

Dippers sing throughout the year except during moult in July and August, and when they are feeding nestlings (Penot, 1948). Song is especially vigorous though in September and October, when birds are establishing winter territories (Hewson, 1967; Galbraith in: Cramp, 1988), and from January to March when they are forming breeding territories (Keicher, 1983; Galbraith in: Cramp, 1988).

Both sexes sing, some observers claiming equally (Eggbrecht, 1937; Witherby *et al.*, 1948; Galbraith in: Cramp, 1988), but others that the male is the more vocal (Keicher, 1983). Our own observations indicate that once a pair starts breeding, it is the male who sings the most. He often sings at intervals during nest-building and when the female is incubating. The female may sing when she is off the nest and her mate approaches, but otherwise she is more silent at this time. Singing occurs at any time of the day although Eggbrecht (1937) recorded song mostly in the morning and evening and less at midday. During the winter though song in mid-morning is commonly heard.

The song has been variously described as somewhat lark-like but subdued (Forrest, 1907); as sweet and thrush-like (Rankin and Rankin, 1940); as a sweet, rippling trill, recalling that of a Wren (Cramp, 1988), and in the American Dipper as nightingale-like but with a more rapid tempo (Ehinger, 1930). Rankin and Rankin noted that the song of the male includes a variety of notes in any order with short phrases repeated and single units. They found that the volume of the song increased when a female approached. By contrast, the female's song is less sweet, being a series of whistles and disconnected units. Shaw (cited in Cramp, 1988) found the female's song less melodious and more scratchy than that of the male. Rankin and Rankin recorded either bird singing alone or as part of a display, but both usually singing as they approached each other. Song bouts lasted up to 6 minutes. Singing often took place from a rock or log in the river, but also occurred on the wing when both members of the pair were together.

Pair-bonding Behaviour

Creutz (1966) found that pair-bonding behaviour was inconspicuous, or almost absent in those pairs which remained together during the winter on their previous breeding territory. It is more marked when a new pair is established; this usually occurs in the spring, but pairs can be formed at any time of the year (Richter, 1953; Fuchs, 1970). Pair-bonding behaviour occurs again once the first brood has fledged and in the autumn after moult (Creutz, 1966), or even on winter territories (Diesselhorst, 1938; Balat, 1962a; Creutz, 1966).

Song is important in attracting a female to a territory. Then one bird, usually the male, approaches the female, who at first moves away from him. Later she may approach him, whereupon he responds by running or swimming towards her, singing as he does so (Creutz, 1966). He may then run or swim around his prospective mate, dipping and blinking frequently, sometimes quivering his wings and making rattling calls, these actions forming the 'advertising display'.

The rattling calls were described by Rankin and Rankin (1940) as a loud rattling 'r-r-r-r' or a short, low, rolling 'zur-r-r-r'. They are commonly given when members of a pair greet one another, but earlier in the season are used in courtship when the female is soliciting, or by the male when he is forming a pair-bond.

Threat and Advertising Displays

Early in the season, the male may behave quite aggressively to the female, performing threat displays towards her with fanned tail, and pursuing her, or bill-fencing with her. During advertising displays, the male may flick his tail up and down, whilst the female frequently has her tail raised and spread (Manuel, 1949; Sudhaus, 1972). Advertising displays take a variety of forms, well described by Eggbrecht (1937) and by Efteland and Kyllingstad (pers. comm.).

Ground displays take the form of a pair facing each other, their bills almost touching, and both singing in a duet. One, or sometimes both birds, extends its head, bill pointing upwards, showing off its white breast. Sometimes the head may be turned slowly from side to side, the tail raised and lowered, the wings half-spread, either stationary or quivering. The male may also perform a bowing action, raising and lowering his head and vibrating his wings, while the female sings. Depending on the situation, the other bird may crouch low over the rock,

Dippers bill-fencing

Bill-fencing between mates: ground advertising display with bills of male and female touching

Ground advertising display with male vibrating wings as female sings

wings fluttering. Rankin and Rankin (1940) found that generally one bird displayed while the other sang, the whole performance only lasting 10–15 seconds. Very similar displays take place between rival birds, so it is not always easy to decide whether such displaying birds are a pair or not, but the resulting behaviour of one or both birds may throw light on this.

Pursuits of the female by the male may lead to high flight displays; song and display flights high over a river during courtship are not uncommon. Both birds fly close together, one behind the other, both singing continuously. Moody (1955) observed such a display flight in Yorkshire, but where only one bird sang. The pair of Dippers were displaying and singing to each other by the river edge when suddenly they flew downstream and then rose to a considerable height, flying round twice together in a wide arc; one bird flew ahead of the other, the trailing bird singing continuously. One of them then completed two similar circuits alone, still singing. A similar display flight was observed by Robson (1956) but that occurred when the pair had nestlings. After a display flight, the birds sometimes land in the water and swim to the edge, to begin displaying again (Hewson 1967; pers. obs.).

Soliciting Displays

As the female becomes more receptive, she approaches the male more and performs either advertising displays or soliciting displays. In the latter she crouches in a submissive posture, spreads and quivers her wings, often begging for food, singing or making rattling calls. The male may respond by singing or rattling to her, or offering her food. Once the male freely offers the female food

Ground advertising display with female singing and male bowing

Aerial display flight: male and female

Soliciting display: female crouching and male singing

and she accepts it, pair formation is complete (Creutz, 1966). Rankin and Rankin (1940) describe seeing on several occasions a bird try to snatch food from the other, with a tug-of-war ensuing, this being accompanied by song, and ending by one bird raising and vibrating its wings, falling backwards into the water or on to another stone where the display and song was sometimes continued.

MATING

Mating occurs anywhere within a pair's territory and at any time of the day. The male may at first ignore the soliciting female, or she may rebuff him, lungeing at him with raised wings (Creutz, 1966). She may simply be uncooperative and turn towards him rather than away, but eventually the male is able to mount, jumping or flying on to her back. The birds maintain their balance during copulation by beating their wings (Eggbrecht, 1937; Creutz, 1966). After the brief copulation of a few seconds, the two birds may fly away from each other, often with a burst of song, or stand around dipping, blinking and preening or begin feeding together.

POLYGYNY

The mating system of the Dipper is usually monogamous, both members of a pair helping in nest-building and with care of the young, and often remaining together through the winter and into the next breeding season. Fidelity to mate from year to year in monogamous pairs is common. For example, one long-lived pair that we first found breeding in 1984, nested together at the same site in each successive season until 1990 after which they both disappeared, presumably dying.

Despite widespread monogamy and mate fidelity, polygyny where a male assists in feeding two or even three broods generally in contiguous territories up to 800 m apart, has been recorded in Dippers by many observers (e.g. Mork, 1975; Galbraith, 1979; A. John, pers. comm., in Devon; C. Mathews and J. Arnold, pers. comm., in Avon;). Sacher (1980) observed two polygynous males attending nests that were asynchronous by one and two weeks, and which were 2 km and 3.5 km apart. Lewis (1991) found one male Dipper overflying a neighbour's territory to feed a second brood, and he also recorded a case involving a male with four females. In Norway occupied nests were recorded only 50 m and 100 m, and once only 8 m, apart suggesting possible polygyny (Efteland, 1983b). In some areas males outnumber females in the breeding season (Richter 1954a; Mork, 1975); on other rivers females outnumber males. It is in these latter situations that polygyny has usually been noted (C. Williams and J. Arnold, pers. comm., on rivers in Avon; Lewis 1991 on a river on the Gwent/

Brecon border). Polygyny may be a result of the high female ratio, as suggested by Galbraith (1979); it is of interest in this context that the case of the male noted by Lewis (1991) as helping four females at their nests, occurred in a year after two seasons with high first-year mortality, which affected males more adversely than females. Alternatively, polygyny might lead to fewer males.

In our Welsh study area, where nests occurred within 100 m of each other, we had often suspected polygyny. However, the sex ratio is usually close to unity (Ormerod, Tyler and Lewis, 1986), so it would seem unlikely that polygyny could be common. In 1989 a detailed study was made by Sonya Yoerg of Dippers on the lower 5 km of the River Edw in mid-Wales. We knew there to be a high breeding abundance on this stretch, with eight or nine known occupied nests in most years; we also had good data on the sex ratios of young and old birds during the previous autumn and winter, and knew that on this stretch males outnumbered females. Many birds had been metal and colour-ringed so were individually recognisable. Yoerg showed that three or four males were polygynous, with one of these attending three different nests. DNA fingerprinting of blood samples from two broods fed by the same male proved conclusively that he was the father of both broods.

Price and Bock (1973) recorded 4 of 31 male American Dippers in their study area in Colorado to be polygynous, each assisting with the feeding of two broods. They found that these polygynous males took little part in nest construction. The two nests of each male were from 180 to 440 m apart with one bird visiting nests over 3 km apart. The difference in laying dates at the two nests of the polygynous males varied from 14 to 40 days, with a mean of 20.5 days, i.e. usually a 2–3 week difference. Thus the polygynous males would only have had to feed both broods at the same time for at most a week. Price and Bock found that the productivity of polygynous birds was significantly higher than that of monogamous birds.

Several aspects of the biology of dippers are common to some other species which are polygynous. They are sexually dimorphic in size as are, for example, Cetti's Warblers (cf. Bibby, 1982; Bibby and Thomas, 1984); they occur in two-dimensional habitats with territories of variable quality (cf. Verner and Wilson, 1966; Orians, 1969); they also build a domed nest (cf. Haartman, 1969). However, if equal numbers of males and females occur in most dipper populations, then we believe polygyny in this species must be opportunistic rather than typical.

NEST-BUILDING

In Britain and Europe Dippers may start building or repairing nests during January and early February, but bad weather will deter them. Birds take longer to build nests early in the season; they stop frequently and the completed nest may remain empty for two or more weeks.

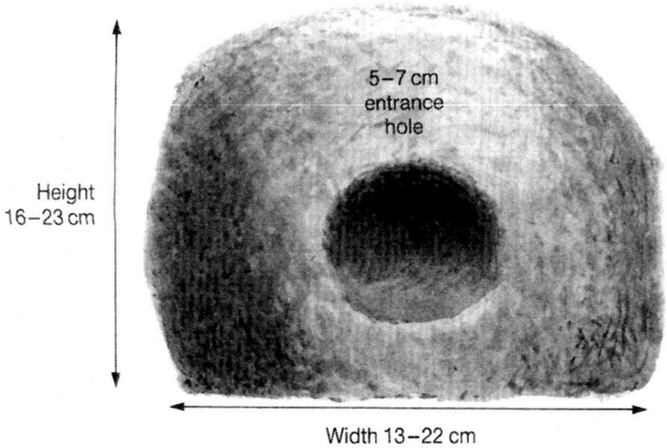

Sketch of nest

The typical finished nest is a large, almost circular, domed structure made mainly from moss with an inner cup of grass, stems and rootlets lined with leaves. The circular entrance hole faces downwards, usually over running water. Where the nest is constructed within a confined space, such as in a narrow drainpipe or crevice in a wall, it may lack part of the dome, even being reduced to just a mossy base and inner cup. Occasionally even where space is not restricted, birds may build an open nest as recorded by Shooter (1969a) in Derbyshire

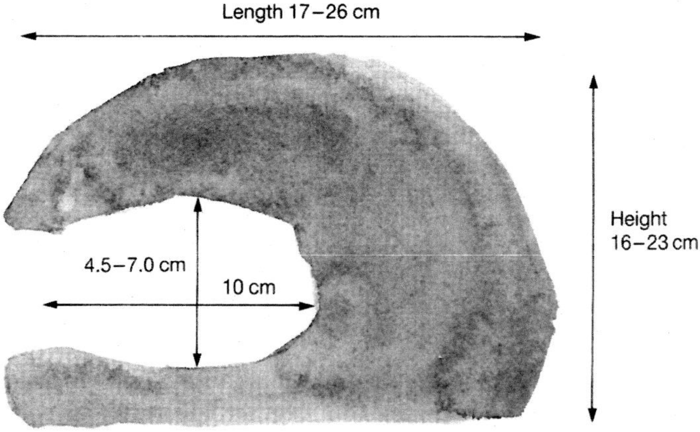

Vertical cross-section of nest

where the eggs were visible from some distance. Efteland (1983b) recorded an extraordinary nest with a 20-cm long funnel-shaped nest entrance overhanging the ledge on which the nest was sited.

Both sexes of a pair investigate a potential nest site but it is uncertain whether it is the male or female who makes the final choice (Shaw, 1978). Rankin and Rankin (1940) found that most nest-building occurred in the early morning. Whilst both sexes usually help build their nest, the female generally completes the cup and lining and sometimes she does most of the nest-building. Fuchs (1970) watched two pairs where the males, both first-year birds, took no part at all, although in subsequent years they did help. Females when building may be visited by their mates, who sing or make rattling contact calls nearby, even when clinging to the nest (Eggbrecht, 1937). Males also frequently pause from working on the nest to sing (Rankin and Rankin, 1940; Manuel, 1949). A division of labour between the male and female of a pair was noted in Yorkshire by Freethy (1980). One bird, probably the male, repeatedly flew with grass and leaves to a stone near the nest. Its mate then collected the nest material, and returned to the nest, into which she incorporated it. Rankin and Rankin (1940) similarly watched a male pass nest material to the female in the nest; in that pair the male made more visits with nest material compared to the female, who stayed in the nest for longer periods, carefully incorporating the moss and other material. During nest-building this male behaved aggressively to the Rankins. He flew towards them in a peculiar upright manner, prominently displaying his white breast. In Wales we have at six or more sites observed similar behaviour by a male when his mate was incubating. At one site he flew directly at us, hovering briefly in front of us.

Nest Structure

Dipper nests are extraordinary structures; they may be large and round, the size of a football, but sometimes lack the dome completely. They are usually made largely of moss with an entrance hole in the side leading into an inner chamber with a lined cup. The sequence of nest construction varies according to the nature of the site. Where the nest is built against a vertical rock face or wall, a circular base for the back of the nest is built first, with the birds building outwards, adding the walls, roof and floor. The inner cup of grasses is added once the basic structure is complete, and finally the leaf lining. To help fix the nest to the rock or wall, birds sometimes initially add mud with the moss. Williams (1984) watched a pair plaster weedy mud on to brickwork with a dibbling bill action, much like that of a House Martin. Once they had completed the circular framework, moss was added to form the walls. This mud base is essential on vertical sites where there is no ledge or crevice to provide a starting point.

At flat sites as on ledges or in drainpipes, the moss base of the nest is built first with birds working upwards and outwards from this, as also observed by Balat

Nest site of a Dipper on a rock ledge

(1964) and Schmid (1985a) in Czechoslovakia and Germany. Sometimes a huge platform may be constructed to provide a base for the nest, which finally appears to be almost rectangular. Mason (1953) described such a nest on a rock as a depressed miniature haystack.

Nest Materials

Nest materials vary according to the availability of different species of moss and grass, although Coward (1969) suggested that Dippers selected materials to help camouflage their nests. In wet situations, the moss in the nest remains fresh and green, further helping disguise the nest. At dry sites, however, the moss soon dries out so that nests become very conspicuous. Such sites though tend to be hidden below bridges.

Some nests at moorland sites in upland Wales are composed largely of Purple Moor Grass, and Mason (1953) described a Yorkshire nest constructed almost entirely from one species of moss. Generally, however, the main structure includes a variety of aquatic or terrestrial mosses or liverworts, common in the locality (Balat, 1964; Hewson, 1967; Perry, 1986). The mosses, gathered from within 30 m or so of the nest site, either from rocks in the stream or from banks and in nearby grassland or woodland, are often wetted before being taken to the nest.

Bryophytes in the Nests of Dippers

Arocladium
Brachythecium rivulare
B. rutabulum
Drepanocladus sp.
Eurhynchium praelongum
E. riparoides
Fontinalis sp
Hypnum cupressiforme
H. cuspidatum
Lophocolea bidentata
Omalia trichomanoides
Radula complanata
Rhytidiadelphus squarrosus
Thamnium alopecurum
Thuidium tamariscinum

The inner cup is woven from grass stems, often Purple Moor Grass at moorland sites, sedges or woodrush. Many authors (e.g. Schmid, 1985a) refer to the preference shown by Dippers for lining the nest cup with dead beech or oak leaves. These leaves remain on the trees long into the winter, rotting only slowly; hence they are available in the spring. Other authors (e.g. Coward and Oldham, 1910; Serle and Bryson, 1935) noted leaves of ivy, bramble and willow, in addition to those of beech and oak, as well as bracken, rush stems or wisps of grass in the lining. Hair and feathers have also been recorded (Watson, 1972) but we consider this very unusual. In our study area in Wales, oak leaves are most numerous, recorded in 47% of nests, with beech leaves in 30%, ivy (11%), bramble (3%), willow (6%) and birch (3%). Generally the lining is composed of leaves of only one species, but there may be a mixture.

Construction Time

The time taken to build a nest varies according to the season and weather but Shaw (1978) gave an average of 28 days for nest construction (range 25–32 days, $n = 40$); in Ireland a period of from only 10 days to more than 21 days was noted (Perry, 1986). In Germany building of 18 nests took from only 9 to 28 days with a mean of 18 (Zang, 1981); whilst Schmid (1985a) recorded building taking from as little as 5 to up to 45 days. One nest in Czechoslovakia was completed in 12 days (Balat, 1964).

Cold or wet weather can delay building early in the season. Thus early nests might take a month to be completed but later nests only 9–14 days (Hewson, 1967). Similarly, where time is limited because the breeding season is very short as at high altitudes, or after the failure of a first nest, birds build more quickly. Breitenmoser-Wursten (1988) recorded 6–10 days for nest-building after a failure, compared with 9–39 days for first nests.

Once a nest is completed, birds may lay immediately or the nest may remain empty for 2–3 weeks. Sometimes a pair abandons its new nest, and relines an old one in which the clutch is laid.

NEST SITES

Natural Sites

Nests are almost invariably built by or over running water. The most usual natural site is on a rock ledge by a river (Table 9), sometimes close to or even behind a waterfall. Other favoured sites are in or on earth and rock banks, often below an overhang, some entirely concealed in crevices. An upturned tree may also provide sites in its roots, and less commonly nests may be on rocks in mid-river, on fallen trunks and branches across a river, or concealed in holes in trees close to the river. Tree sites are, however, only rarely recorded (Serle and Bryson, 1935; Robson, 1956). Shaw (1978) found none listed on 1159 BTO nest record cards. Shimeld (1984) recorded one nest on the horizontal branch of an alder, 2.5 m over the water; Younger found a nest in a fork of a tree more than 3.5 m above the ground (Bannerman, 1954), whilst Watson (1972) described sites in ivy on trunks and branches. Trochet (1967) also recorded a Dipper nest in France on a tree. The late Colonel Morrey Salmon (pers. comm.) found a nest in mid-Wales in a hole in a large branch of a sycamore that had fallen above a waterfall. In our study area in Wales only 2–6% of nests found in any one year are in or on trees. We have found nests on branches and trunks fallen over streams, and in holes in alders or wych elms by rivers, some of these sites being used regularly. The two most unusual tree nests were suspended in flood debris in the tip of branches overhanging the water.

Artificial Nest Sites

The availability of good nest sites must determine a pair's choice of where to build. In remoter upland rivers where rocks and cliffs abound but where there are few bridges or walls, most nests are built on rock faces or in banks; for example, 87% of nests found by Robson (1956) were in natural sites (Table 9).

Table 9. Variation in choice of nest sites by Dippers in different study areas

Source	No. of sites	Natural nest sites (%)							Artificial nest sites (%)					
		In/on rocks/ rock-faces	*Waterfall*	*In bank*	*Tree root*	*In/on tree*	*In vegetation*	*Total*	*Ledge/ crevice in/on/ below bridge*	*In wall (not under bridge)*	*Pipe/ culvert*	*Nest-box*	*Other*	*Total*
Shaw (1978)	1159	24.50	5.60	10.40	6.40	2.80	0.90	50.60	34.50	6.90	3.10	–	4.50	49.40
Wales 1986/87 (our study)	252	14.30	3.20	4.50	4.20	2.00	–	28.20	44.40	21.40	4.70	0.40	0.80	71.70
Robson (1956)	50	64.00	6.00	4.00	–	4.00	2.00	90.00	14.00	4.00	2.00	–	–	20.00
Hewson (1967)	12	8.30	–	–	–	–	–	8.30	83.30	8.30	–	–	–	91.70
Perry (1983)	41	–	–	10.00	5.00	–	–	15.00	78.00	2.00	5.00	–	–	85.00
Balat (1964)	66	34.80	–	–	10.60	1.50	–	46.90	33.30	3.00	–	16.60	–	52.90
Andersson and Wester (1975)	84	56.00	–	15.50	–	–	–	71.50	?	?	?	?	?	28.50
Efteland and Kylingstad (1984)	60	41.00	14*	–	–	–	–	55.00	45.00	–	–	–	–	45.00
Schmid (1985a)	180		4.40	–	4.40	–	0.60	9.40	11.10	14.50	–	65.00	–	90.60
Breitenmoser-Wursten (1988)	77	19.50	6.50	9.10	–	1.30	–	36.40	18.20	25.90	–	19.50	–	63.70
Czapulak (1988)	91	12.00	–	–	5.50	–	–	17.50	74.80	7.70	–	–	–	82.50

NOTES:

Some of the categories are ambiguous. For example, some sites under walls or pipes could be in or under bridges; most nest-box sites are also under bridges.
Efteland and Kyllingstad (1984) included waterfall sites with other sites as in buildings.

Unusual nest sites of a Dipper in flood debris in branches overhanging a river (nest incomplete)

Where old walls, stone bridges or girder bridges are present, birds often use these rather than natural sites. In our Welsh study area some 72% of 252 nests were in artificial sites, notably ledges on stonework often associated with bridges, girders below bridges, and crevices or pipes (Table 9). In Germany, Schmid (1985a) found that as many as 90% of nests were in man-made sites. Many other observers note the use of bridges as Dipper nest sites (e.g. Balat, 1964; Shaw, 1978; Perry, 1983).

Bridge and wall sites rarely collapse as do banks and cliffs, and most bridge sites are safer from flooding than natural sites. At one site where birds nested on a ledge below a farm bridge for four or more years in succession, they moved to a rock face downstream only when the bridge was reinforced and the ledges eliminated. On some territories birds alternate between natural and bridge sites.

Although there may be a bias towards observers finding nests at bridge sites, and in subsequent years rechecking such sites, our intensive study in Wales does suggest that where Dippers have a choice, they do prefer bridges rather than natural sites. In Switzerland, Breitenmoser-Wursten (1988) found that bridge sites allowed birds to build nests and lay eggs earlier than at natural sites. Shaw (1978) found, however, that the incidence of successful nests, defined as those which produced at least one fledged young, was similar at natural (62.5% success) and artificial sites (60.5%). On some more lowland rivers where the gradient may still provide good Dipper habitat, there may be little choice for nest sites apart from bridges; these have perhaps enabled Dippers to spread further from their upland strongholds.

Fidelity to Nest Site

The great majority of nest sites, particularly bridges, are re-used in successive seasons, although on some territories birds may use two or three different sites in different seasons or even in the same season.

Dippers building under bridges often prefer a particular ledge. For example, a bridge in the Welsh Borders with six ledges on each side has been occupied by Dippers for at least 14 successive years. In 11 seasons, they built on an outermost ledge on the downstream side of the bridge; in just three they chose a ledge on the other side of the bridge. At many other bridges, exactly the same site is used in successive seasons, although sometimes, perhaps following a nest failure or change of partners, a nest may be built on a different ledge.

A pair may build alongside an old nest (Schuster, 1953; Sudhaus, 1972; Welsh study) rather than remove or repair it, but generally for successive clutches in the same season (Shaw, 1978) and even in successive seasons, nests are repaired and relined. In one year Robson (1956) found that 14 of 25 nests from the previous season were re-used.

The use of traditional sites by Dippers is well known and many authors have commented on birds using the same site for 30 or more years (Morris, 1896; Morrey Salmon, pers. comm.), with the longest record 123 years (Gladstone, 1910).

Nest-boxes

Open-fronted nest-boxes have been used successfully for Dippers throughout Britain and Europe, particularly in Germany (and for American Dippers in the United States). Jost (1970), for example, experimented widely with two designs of boxes, both of which were readily accepted. Nest-boxes can be used to retain breeding pairs at sites which are made unsuitable, such as where old bridges are replaced with concrete box bridges, to try to prevent a further decline in populations, as in Hungary (Horvath, 1987), or to increase populations on rivers where few apparent nest sites are available. In one area of Germany, the provision of nest-boxes made from wood and bark and attached to trees, increased the number of breeding pairs on a stream from 1 to 14 pairs (Staedtler and Bremshey, 1988). Chapter 10 provides further information on nest boxes.

Height of Nests Above Water and Their Distance From Water

Most nests are built by running water, sites even a few metres from water being unusual (e.g. Andersson and Wester, 1975). In dry springs, though, as in 1989 and 1990 in Wales, nests by late May may be 2–3 m from the contracted stream. The very few records of sites some distance from water include a Norwegian nest built in a wood 30 m away from a river (Efteland, 1983b) and a nest found by Moon (1923) in the wall of an old byre in Westmoreland 100 m from a stream.

Occasionally Dippers nest in buildings close to water. Taylor (1929) recorded a nest exposed on the roof of a water-wheel house, the nest entrance facing towards the ground, rather than over the stream. In Denmark a pair of Dippers used the middle of three House Martin nests 7 m up in the apex of the roof of an old water-mill (Nyhlen, 1952). Another unusual site was found by Greenhalgh (1969) 3 m down the shaft of a pot-hole in Ribblesdale in Yorkshire. The young birds would have been faced with a 70-m drop down the shaft or a vertical flight up and out of the shaft. In Gwent a pair of Dippers nests each year in a ledge in a wall behind a working water-wheel. Most nests are built less than 3 m above water (Robson, 1956; Mawby, 1961; Shaw, 1978), although clearly the height of a river may rise and fall up to a metre or more during the breeding season. Shaw noted that only 2.4% of 1276 nests were above 5 m. In Wales we have also found that 91% of nests are below 3 m, with over 60% between 1 and 2 metres, and only 2% over 4 m. On small tributaries nests have been found only a few cm above water.

Dipper nest-box with the nest of an American Dipper in California

In Germany, Dick and Sackl (1985) found that height of nest sites above water level and various other factors such as width and flow of the river showed great variation and had no influence on nest-site selection.

Multiple Nest-building

Where similar nest sites occur side by side, many species of bird may build two or more adjacent nests. Such multiple nest-building has been reported in Grey Wagtails (Tyler, 1972). Ledges and girders below bridges provide a series of almost identical potential nest sites and occasionally a pair of Dippers starts building two or even three nests on adjacent ledges. Only rarely do they complete both, or lay eggs in both. Shooter (1969b) recorded two birds (presumably a pair) building separate nests 45 cm apart in the roots of a fallen willow in Staffordshire.

NESTS OF OTHER SPECIES OF DIPPER

All other species of dipper construct nests with a similar design and in a similar range of sites to the Eurasian species. Thus the structure of nests of the American

Dipper appears to be identical to that of the Eurasian Dipper, having an outer 'shell' of moss with small amounts of grass, an inner cup of grass stems and a lining of dry cottonwood leaves (Bakus, 1959a), or of dry needles of Ponderosa pine (Goodge, 1959). Muir (1894) poetically describes a nest as resembling an old-fashioned brick oven or Hottentot's hut. He noted the species of moss used in the structure was plentiful on rocks and logs near waterfalls. Bent (1948), quoting Junius Henderson (1908), said that nest-building was performed mainly, if not wholly, by the female. The American Dipper also re-uses nests from year to year, adding more moss, a new cup and making the nest entrance smaller (Bakus, 1959a).

An identical range of nest sites is used by the American Dipper as by Eurasian Dippers. For example, Bent (1948) wrote that nests were sometimes below a waterfall, or on a rock in mid-stream, but more often on a narrow ledge on a wet rock cliff among mosses and ferns. He also recorded nests built among upturned roots of fallen trees, in a cavity in a stump, in crevices, on girders or beams under bridges, and one in a wooden shed above running water. Bakus (1959a), too, recorded nests below bridges and another over a dam spillway. As in Britain, the location of the nest site is determined by the habitat and availability of sites (Johnston, 1943).

Hann (1950) believed American Dipper nests had two characteristics in common: a location over or by water and a high degree of inaccessibility. Certainly nests away from water are rare but one nest was reported over 9 m from water on a sloping roof of a building (Sullivan, 1966). Nest sites of American Dippers are generally at greater heights than those of their Eurasian relatives, probably because river levels fluctuate more widely in the Rockies than in Britain and Europe. Bakus (1959a) found 13 nests from 2 to 6.5 m (mean of over 3.5 m) above water. Price and Bock (1983), in their long-term study in the Rocky Mountains in the Front Range of Colorado, concluded that the quality of nest sites was one of six major factors affecting the American Dipper population in the Boulder area. They ranked quality on four criteria: height above water; width of rock ledge; presence of an overhang; and security from predators. High, inaccessible, wide ledges with an overhang were considered the highest quality nests because these were safest from predation or flooding, the latter being a major cause of nest failure. Bridge sites were considered to be excellent. The quality of nest sites was determined, therefore, by both geology and man's activities.

Less is known of the nests of the other three species, but from the accounts available, and from our own limited observations of these species, it is evident that the structure and locations of nests of the Brown, White-capped and Rufous-throated Dippers are remarkably similar to those of dippers elsewhere. Thus the nest of the Brown Dipper is 'a massive globular structure of green moss and dried grass lined with dry leaves and moss roots, placed in clefts of moss- or fern-covered rocks, on a ledge of cliffs or among boulders — sometimes in a hole in a fallen log or in a hollow in the masonry of a bridge — usually within a metre from water. Occasionally behind a waterfall or cascade. When built in holes, the structure may or may not be domed' (Ali and Ripley, 1983).

Sunquist (1976) observed that moss was the main component of eight nests of Brown Dippers in Pakistan. This was gathered from rocks along the river, occasionally dipped in the water and taken by both birds of a pair to the nest; it was not possible, though, to ascertain whether both birds built the nest. All three visible nests were on rock faces 0.3–2.0 m above the water, and appeared as large balls of moss of very similar dimensions to those of the Eurasian Dipper. Other nests were concealed under rock piles in or by the river. The four Brown Dipper nests that we saw in Nepal, in both their appearance and their locations in rock crevices, on rock faces by waterfalls and under a rock overhang on a mid-river boulder, closely resembled those of our Welsh birds.

Zarudnyi in 1912 (quoted in Kovshar, 1979) commented on the similarities of nest structure and sites of Brown and the white-bellied race of the Eurasian Dippers in the USSR, but other authors such as Stepanyan (1962) noted differences in nest locations. Brown Dippers nested on cliffs over deep water, whilst the white-bellied Dippers nested under low river banks in dense overhanging moss. Budris (1981b) also found a nest of white-bellied Dippers in a small hole under an overhanging layer of peat, whereas Brown Dipper nests that he found were on crevices between rocks on the bank or in mid-stream. Kovshar (1979) suggested that Brown Dippers in the subalpine region of Tien Shan preferred cliff faces, only rarely nesting behind waterfalls, which he found to be favoured by white-bellied Dippers. All these studies were limited by very small sample sizes. It seems likely that the Brown Dipper, as the Eurasian and American Dippers, will use a range of sites depending on availability.

A nest of the White-capped Dipper from Ecuador, found by Sergio Lasso, was below a small waterfall, built into a cleft on a moss-covered rock cliff, 1.3 m above a small rocky river. It was roughly spherical with a diameter of 25 cm, and consisted of an outer moss layer with a circular entrance hole, and an inner cup mainly of dry leaves, strips of bark and strips of plastic. The diameter of the entrance hole was almost 8 cm but as the nest contained nestlings when first found, the entrance hole had become enlarged. Other old nests on the river were in exactly those sites which a Eurasian Dipper might have chosen.

In Argentina, Olrog (Fraga and Narosky, 1985) first observed nests of Rufous-throated Dippers. Salvador, Fraga and Narosky (1986) then described three in niches in cliffs by the Los Sosa River in Tucuman province at heights of 0.95–1.25 m above water. Their descriptions again resemble those of the nests of other species of dipper. The nests were bulky globular structures of green moss with a side entrance, measuring 23–24 cm in height, 22 cm in width, and 14 cm in depth, and the entrance holes 5–7 cm in diameter. A deserted nest removed for examination had two well-defined parts: an outer globular shell and an inner cup. The outer part comprised two species of green moss, grass stems, a few rootlets and pieces of filamentous green algae. The inner cup was built with grass stems, some green algae and leaves of Andean alders, with pieces of paper and

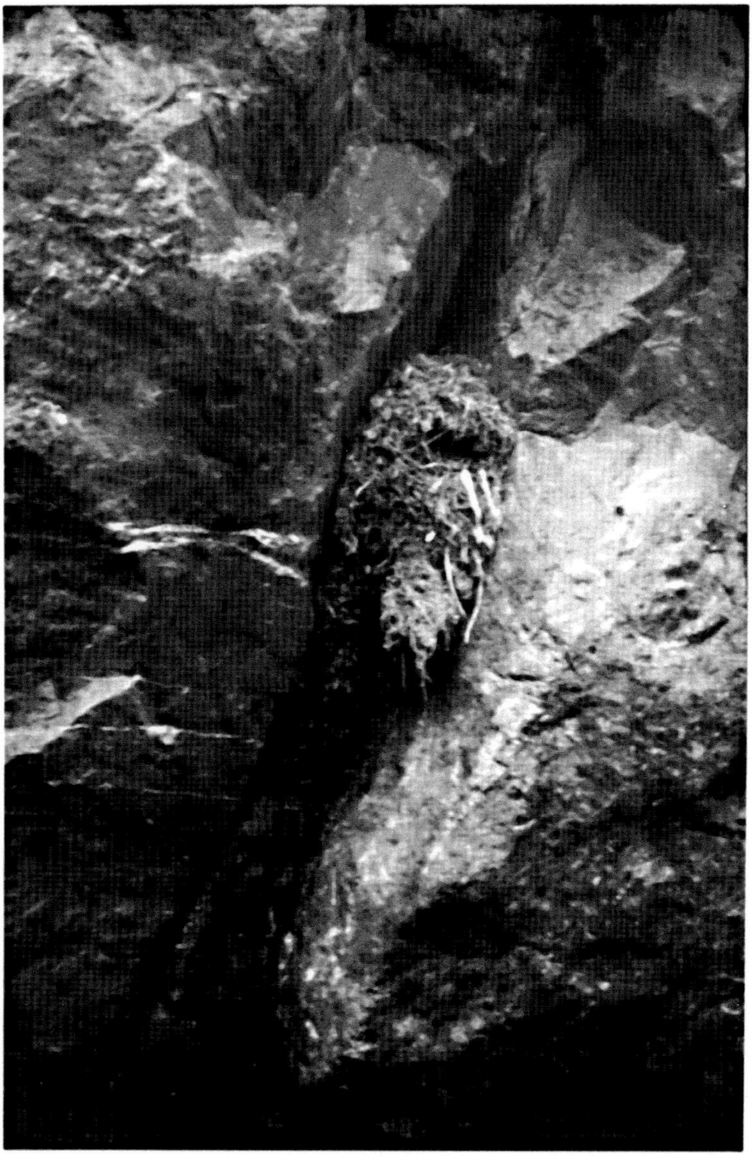

Nest of Rufous-throated Dipper in Argentina

plastic in the lining. Five nests that we found in Tucuman province in October 1993 were remarkably similar to those of Dippers in Wales and were in identical sites — over deep water on the wall of a bridge, on rock cliffs below overhangs or over a waterfall and in a boulder bank. Two were still being built and at one of these an adult, presumed to be the female, spent much time taking material into the nest and incorporating it into the sides of the dome. Her mate in the meanwhile sang from a mid-river boulder. He accompanied her as she flew to and from the nest. None of the five nests that we found had litter incorporated into the structure but all except one were well away from roads and associated human debris. Only rarely do bits of polythene and paper become incorporated into Dipper nests in Britain and Europe, though in a small number of nests in Wales we have seen such litter as well as bits of fishing-line, presumably gathered with moss from rocks in the river.

CHAPTER 6

The breeding season

In this chapter we provide information on egg-laying, clutch and brood size, and on the behaviour of dippers during the incubation and fledging periods; we also look at factors which affect the time of laying and the size of clutches and broods, and discuss the adaptive significance of the early breeding season and of other described patterns. Much of the data refer to the Dipper in Europe, but reference is made to other species of dipper where relevant and where known.

TIME OF LAYING

Dippers are amongst the first species to start laying in the early spring. Shaw (1978) in his analysis of BTO nest record cards found that Dippers in Britain laid eggs from late February until the second week of June, with the peak laying period in late March and early April (Fig. 40). Over half of all clutches were started by the middle of April. In 364 clutches laid between 1978 and 1983 in south-east Wales, we found that there was a peak of egg laying in the first week of April with 52% of clutches started before 7 April (Tyler and Ormerod, 1985).

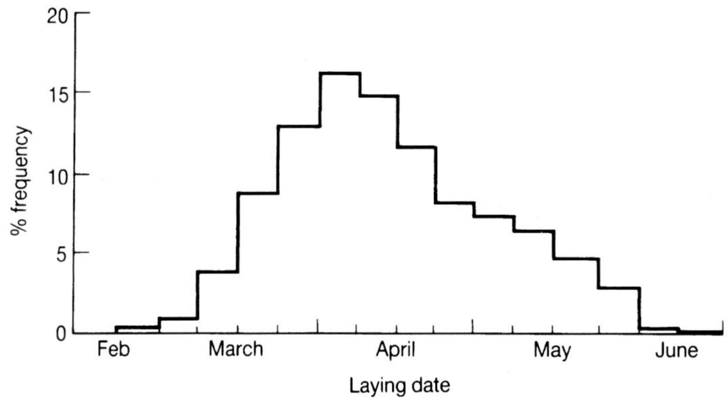

Figure 40. The breeding season of Dippers in Britain as shown by the percentage frequency distributions of laying dates of the first egg of clutches (Shaw, 1978).

From a larger sample of 1263 clutches laid between 1978 and 1993 in our wider study area, a peak in laying was evident in late March (Fig. 41), as found by Shaw. However, unlike Shaw's pattern, our south-east Wales sample and our larger sample both showed a clear second peak of laying in early May. Studies by Baines (pers. comm.) and Robson (1956) also showed a second peak in laying and they gave laying dates which generally did not differ significantly from our own.

February clutches are rare. Thus only seven (0.5%) of 1263 clutches in our study were initiated before the beginning of March. Jourdain (1911) recorded an exceptionally early laying in north Wales with a Dipper on a clutch of eggs on 10 February. In Austria and Germany earliest clutches were in early to mid-March (Schmid, 1985a; Wagner, 1985; Kaiser, 1988; Sackl and Dick, 1988),

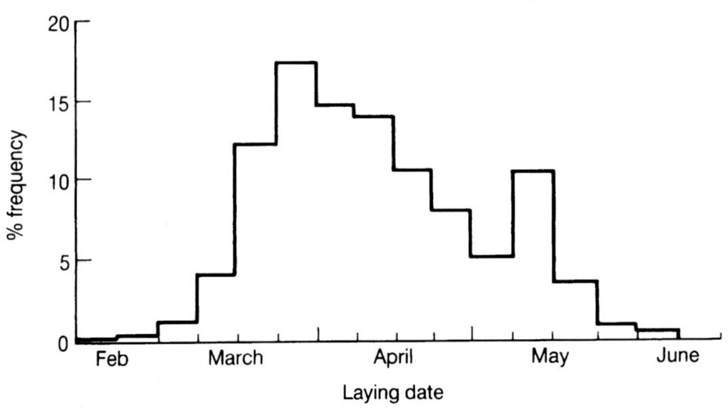

Figure 41. The breeding season of Dippers in Wales between 1977 and 1993 as shown by the percentage frequency distributions of laying dates of the first egg of 1263 clutches.

whilst in Norway and Finland, Haartman (1969) noted that clutches are not laid until early May in the south and mid-May in the north. Efteland (1975), however, recorded April clutches as normal in south-west Norway. In North Africa laying occurs from mid-March (Heim de Balsac and Mayoud, 1962).

A bias towards observers only searching for nests of passerines during the main breeding season is believed to account for a general paucity of nest record cards from later in the summer. This is unlikely to be the case in the Dipper because the second half of its breeding season coincides with the period that most observers are active (May and June) and in Wales, despite our continued searches for second or repeat clutches, we have found that very few eggs are laid after the end of May. Up to 1993 we found only four (0.3%) clutches started in June (Fig. 41), two of these occurring in 1979, one in 1986 and one in 1993. The nestlings from these late clutches would not fledge until towards the end of July, at a time when most adults would already be well into moult.

Factors Affecting Time of Laying

Low temperatures and heavy rainfall in the late winter appear to delay laying. In years with severe weather in January and February, laying dates were significantly later than in other years (Tyler and Ormerod, 1985). Conversely, mild winters appear to advance laying as our few February clutches have been in years following relatively high winter temperatures. For example, following the exceptionally mild winter of 1988/89, several females were incubating from 20 February, which indicates laying of their first egg on or before 16 February. In south-west Norway, where April clutches are normal, Efteland (1975) recorded March clutches following mild weather in late winter and early spring. Sackl and Dick (1988) and Schmid (1985a) showed that in Austria and Germany, laying dates were affected not only by March air temperatures, but also by water temperatures and by the height of the rivers in April.

Laying is delayed, too, by increasing altitudes. In Wales we have found that clutches above 300 m are laid significantly later than those below 100 m. In Czechoslovakia Balat (1964) found a delay of 5.92 days for every 100 m increase in altitude; in Austria and Germany Sackl and Dick (1988) and Schmid (1985a), respectively, found a delay of 5.27 and 4.98 days per 100 m, whilst in the Swiss Alps Breitenmoser-Wursten (1988) recorded as much as an 18.9-day delay per 100 m. Altitude and latitude are closely correlated with temperature, late winter temperatures being colder at both increasing altitudes and latitudes. Dippers in the highest areas or in the most northern latitudes may therefore be expected to have the latest breeding seasons. This is borne out by the late seasons shown in northern Scandinavia and high in the Swiss Alps. Surprisingly, Shaw (1978) found no evidence in Britain for regional variation in laying dates with latitude or altitude, but his sample was heavily biassed to northern England and Scotland (72% of cards from these areas), and he did note that in south-west England a greater proportion of clutches (50%) were started by the end of the first quarter of April.

Water quality and food availability also affect the time of laying. Dippers at acidic sites in upland Wales, where invertebrate prey and calcium are limited, lay significantly later (by 13–22 days) than birds at good quality sites (Fig. 42) even after removing the effect of altitude (Ormerod and Tyler, 1987b; Ormerod *et al.*, 1991). Early clutches were absent below a pH of 6. Vickery (1991) also found delayed laying at acidic sites in Scotland.

Reasons for Early Laying in Dippers

The early breeding season of the Dipper was interpreted by Shaw (1978) as a synchronisation between the nestling period and the period of peak biomass in aquatic benthic invertebrates prior to the emergence of adult insects. It is of note that in the Welsh Wye catchment some insect larvae such as hydropsychid caddis, which are important prey for breeding Dippers (Ormerod, 1985a), reach their peak individual weight during April to June prior to emergence (Edwards and Brooker, 1982). Other possible influences on the Dipper breeding season such as the seasonal reduction in spates, which occurs in the River Wye in March to April (Ward, 1968), should not be overlooked.

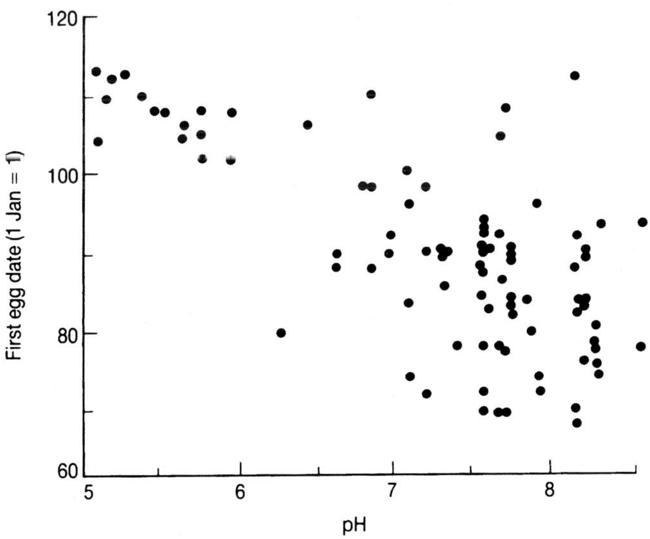

Figure 42. Laying dates i.e. first egg dates, of Dippers in Wales (excluding known relays and second clutches) in relation to mean pH.
($r = 0.468$; $P < 0.001$ (Ormerod and Tyler, 1987a).

Time of Laying in Other Species of Dippers

Other species of dipper also breed early in the year. Thus the west Himalayan race *tenuirostris* of the Brown Dipper breeds from December through to August, depending on altitude, and the eastern race *dorjei* breeds mainly in December and January, occasionally to July (Ali and Ripley, 1983). In Japan Hosono (1978) also recorded Brown Dippers as laying between January and February.

The effect of altitude on laying is apparent from our observations of Brown Dippers in Nepal. There we found birds breeding in the Middle Hills, at less than 1000 m in altitude, in November and December; at around 2000 m they were feeding young in the nest in mid- to late March, whereas at 3000–3500 m they were still building their nests at this time.

Bakus (1959a) noted that the American Dipper laid from late March to July, and Goodge (1959) found that this species laid earlier at lower elevations. Data are few for the South American species, but the limited observations again suggest that the two species of dipper in the Andes start laying before most other passerines. We found nests of Rufous-throated Dippers with eggs and almost fully grown nestlings in Argentina in October, when many other species of bird were still actively courting or nest-building.

EGGS — THEIR SHAPE AND WEIGHT

Birds that nest in enclosed sites usually lay white eggs and dippers are no exception. Eggs of all five species of dipper are white, and are sub-elliptical, smooth and glossy. Egg weights and sizes differ slightly between the species, differences reflecting the size and weight of the species. Thus eggs of the Brown Dipper are the largest (Ali and Ripley, 1983), and those of the South American species the smallest (Salvador, Narosky and Fraga, 1986). Eggs of the (White-throated) Dipper measure *c.* 26 mm × 18.5 mm, and weigh 4–5 g (see also Bent, 1948; Schonwetter, 1971; Ali and Ripley, 1983; Cramp, 1988). Egg shape varies within a species, some Dipper eggs being markedly rounded (e.g. 24 mm × 19 mm) and others more pointed (e.g. 28.5 mm × 19.1 mm).

In Wales and Scotland, eggs tend to be thinner shelled and to weigh less at acidic sites where calcium and invertebrate prey are limited (Ormerod *et al.*, 1988a).

CLUTCH SIZE

In Britain Shaw (1978) found that the Dipper's clutch size varied from 1 to 7 (Fig. 43). However, the smaller clutch sizes of 1 or 2 eggs and the largest of 7 eggs

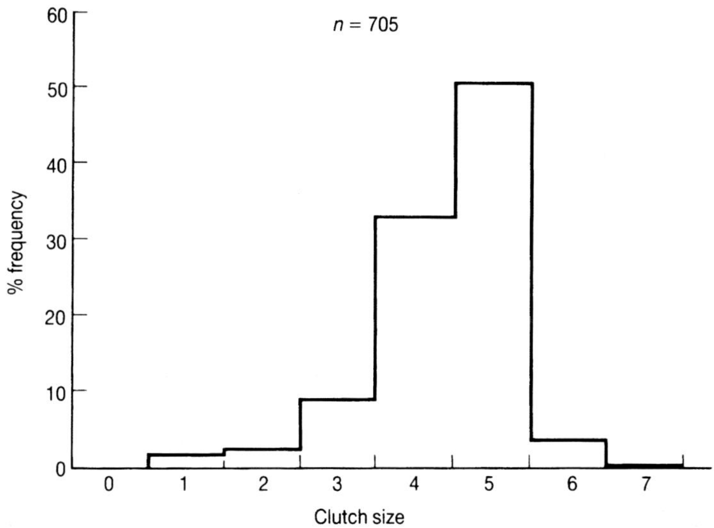

Figure 43. Percentage frequency distribution of clutches of different sizes (Shaw, 1978).

were rare, together comprising only 3.4% of 705 clutches (Shaw, 1978). Clutches of 5 predominated, and most others were of 4, with a mean of 4.42 eggs.

In Wales clutch size in the southern Wye catchment varied from 3 to 6 eggs, with a mean of 4.78 (Tyler and Ormerod, 1985), whilst a larger sample of 427 clutches from south-east Wales between 1975 and 1993 gave a rather similar mean of 4.72 eggs. Clutches of 5 were most frequent (67% of all clutches), with clutches of 4 also common (24.3%); only 4% and 4.7% of clutches were of 3 or 6 eggs, respectively. The mean clutch sizes from these studies were significantly higher than that of Shaw's or than the values given by most other studies in Britain (Table 10). Only the mean values of 4.76 and 4.68 given respectively by Mawby (1961) for the Sedbergh area of Yorkshire and by Shaw (1978) for northern Scotland, were similar, although in parts of mainland Europe clutch sizes are similar or larger (Table 10).

Annual and Seasonal Differences in Clutch Size

Overall, the mean clutch sizes do not differ greatly from year to year, although the proportion of clutches of different sizes does vary between years. For example, in south-east Wales the mean size of all clutches and of those laid before the end of April has slightly but significantly declined between 1975 and 1993 (Fig. 44a and b; see also Chapter 8) due to a greater proportion of clutches of four eggs in the later years and fewer clutches of six.

Table 10. Mean clutch size of Dippers in 17 different studies

Regions	*Mean*	*SD*	n	P	*Source*
SW Norway	5.08	0.72	72	0.1 (*t*)	Efteland and Kyllingstad (1984)
Germany	5.02	–	179	–	Rockenbauch (1985b)
Swiss Alps	4.91	0.58	87	–	Breitenmoser-Wursten (1988)
Poland	4.9	–	20	–	Czapulak (1988)
E Germany	4.85	NA	88	NA	Zang (1981)
SE Wales	4.78	0.61	222	–	Tyler and Ormerod (1985)
NW Yorkshire/ Cumbria, England	4.76	NA	65	NA	Mawby (1961)
Czechoslovakia	4.73	NA	46	NA	Balat (1964)
Germany	4.72	–	142	–	Schmid (1985a)
Grampian, Scotland	4.55	0.79	74	0.03 (*t*)	Hardey, Rae and Rae (1978)
Lower Austria (first clutches)	4.49	1.08	49	–	Sackl and Dick (1988)
Britain	4.42	0.88	705	0.001 (*t*)	Shaw (1978)
N Wales	4.4	0.57	25	0.00((*U*)	D. Moss (pers. comm.)
Cumbria, England	4.34	0.99	38	0.03 (*U*)	D. Baines (pers. comm.)
Derbyshire, England	4.31	NA	26	NA	Shooter (1970)
Cumbria, England	4.2	1.01	45	0.001 (*U*)	Robson (1956)
Grampian, Scotland	3.4	NA	9	NA	Hewson (1967)

NOTES:
SD = standard deviation; *n* = sample size; *P* indicates level of the difference between each study and our study in south-east Wales, using *U*-test or *t*-test as indicated.

Within each year, first clutches started in March and April are significantly larger than later clutches which are mainly relays and second clutches (Fig. 45) (Shaw, 1978; Tyler and Ormerod, 1986). Proportionately more clutches of three eggs are laid towards the end of the breeding season. First clutches laid during the peak of the season are also slightly larger than those laid at the very start of the breeding season.

In Europe results of the effect of seasonal changes in clutch size are conflicting. Sackl and Dick (1988) found no noticeable seasonal effect but in Germany Zang (1981) noted a decrease in clutch size of 0.17 eggs per 10 days of the season, with the mean size of later clutches only 4.38 compared with 4.86 in earlier clutches, and Breitenmoser-Wursten (1988) showed that in the Swiss Alps first clutches were larger than repeat (mean 4.86 eggs) or second clutches (mean 4.40). She also found, as in Britain, that the earliest first clutches (mean 4.94) were slightly smaller than later first clutches (mean 5.11).

The adaptive significance of the most common clutch size and of seasonal changes in clutch size is discussed below, under 'Brood size'.

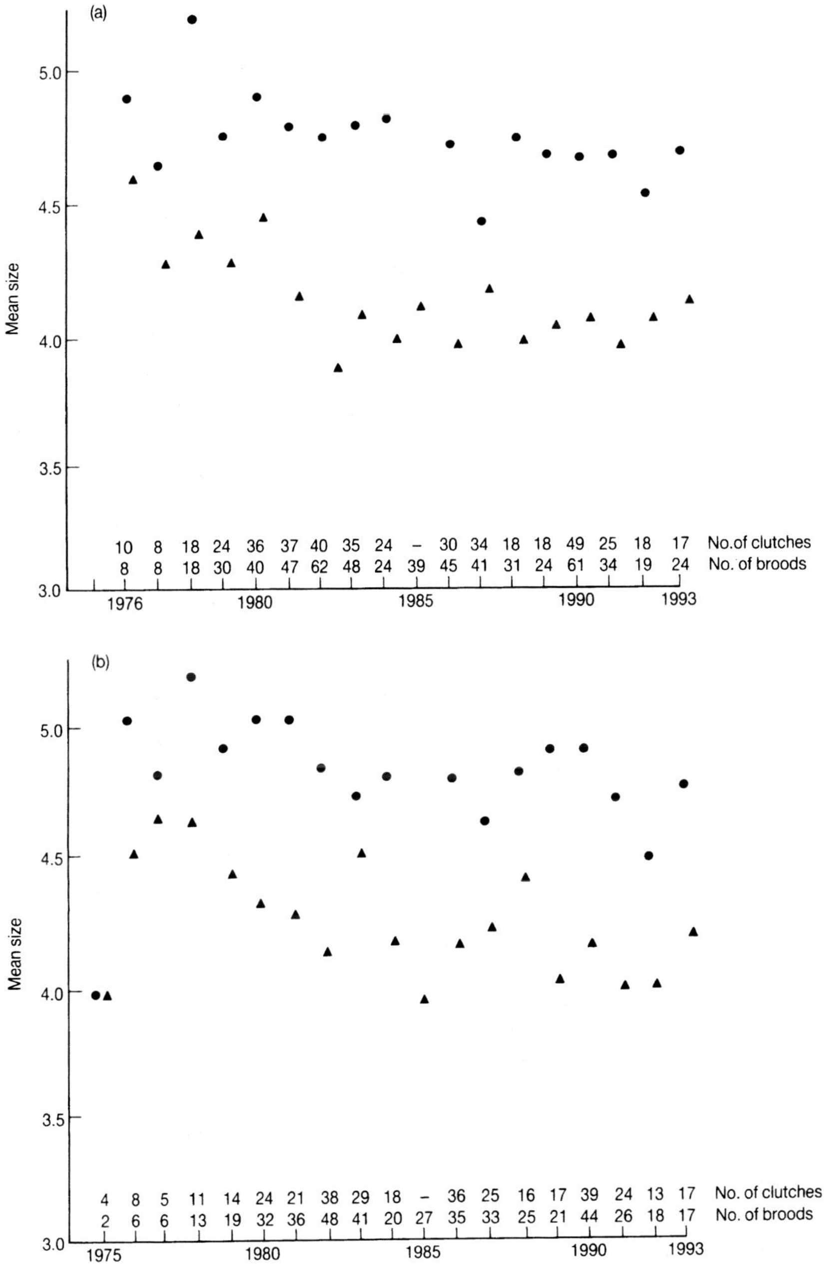

Figure 44. (a) Annual patterns in mean clutch (●) and brood (▲) size in Dippers in south-east Wales. (b) Annual patterns in size of clutches (●) laid before 1 May and of broods (▲) from these earlier clutches.

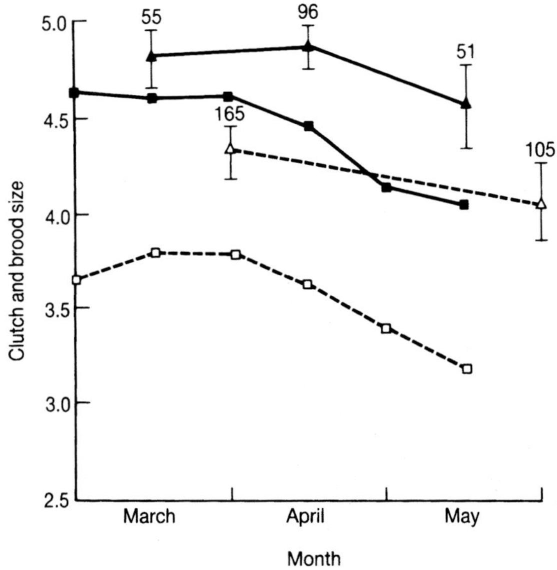

Figure 45. Seasonal patterns in clutch (closed symbols) and brood size (open symbols) from our study in south-east Wales (triangles) and from Shaw (1978) (squares).

Other Factors Affecting Clutch Size

No altitudinal effect on clutch size has been detected in Britain; nor could Sackl and Dick (1988) or Breitenmoser-Wursten (1988) demonstrate any such effect in Austria or Switzerland. Zang (1981), however, reported that clutch size decreased by 0.2 eggs for every 100 m increase in altitude, but this may have been due to later laying at higher altitudes.

The effects of water quality and food supply on clutch size have been investigated in Wales (Ormerod and Tyler, 1987b) and in south-west Scotland (Vickery, 1991) in relation to stream acidity. Birds at poor quality sites (i.e. acidic streams with an impoverished benthic invertebrate community) lay significantly smaller clutches than those at good quality sites. Thus in 1985 and 1986, birds in a wide study area in Wales produced an average of 4.8 eggs per clutch (SD 0.5, $n = 132$) on good quality sites where the pH of the water was greater than 6, compared with only 4.0 eggs (SD 0.9, $n = 25$) on poor quality, acidic sites with a pH of less than 6 (Ormerod *et al.*, 1991). Birds laid later at the latter sites which partly explains the difference, but even where clutches on good quality sites were also laid late, there was still a significant difference. The smaller clutch sizes at acidic sites may be the result of female Dippers incurring problems in finding either sufficient food and/or calcium for breeding (Ormerod *et al.*, 1991).

It is of interest that Zang (1985) found that in the Harz Mountains in

Germany, Dippers nesting on southern facing slopes laid larger clutches than those on north-facing slopes. He related this to the better food supply on the former slopes.

Studies of other passerines have shown strong positive relationships between food availability (Hogstedt, 1981; Perrins and Birkhead, 1983), territory quality (Hogstedt, 1981) and clutch size, although experimental manipulations (providing birds with extra food in the prelaying period) have not always supported the link between the abundance of food and clutch size.

Efteland and Kyllingstad (1984) recorded slightly larger clutches where both parents were adults than when one or both were first-time breeders, but the difference was not significant.

Clutch Size of Other Species of Dipper

Clutches of eastern races of the White-throated Dipper and of the Brown Dipper in India are of 4 or 5 eggs (Ali and Ripley, 1983). In Japan, Asami and Haga (1983) also recorded Brown Dipper clutches of 4 or 5 with a mean of 4.7 eggs. The American Dipper lays 3–6 eggs, usually 4 or 5 (Bent, 1948). Data for the South American species are too few to give meaningful ranges, but our limited observations in Argentina of a clutch and a brood, and the data of Salvador, Narosky and Fraga (1986), suggest that a clutch of 2 eggs is normal for the Rufous-throated Dipper. Such a small clutch size is typical of birds in the tropics; Mountain Wagtails in Ethiopia, for example, usually lay only 2 eggs.

INCUBATION

Dipper eggs are generally laid at one-day intervals although Shaw (1978) recorded a clutch of three laid over five days, and two successive eggs laid with an interval of two days. Not all eggs are laid early in the morning, as we have frequently flushed Dippers from empty nests or incomplete clutches in the late morning or afternoon. Rankin and Rankin (1940) also found that nests that were empty in the morning could have an egg by the evening. During the laying period, the male is very attentive towards the female. The two frequently display and sing to each other, make rattling contact calls, and when the female is on the nest, both during laying and incubation, the male often remains nearby, guarding her.

Most observers have found that it is the female Dipper that incubates the eggs (likewise for the American and Brown Dippers). Balat (1964), however, having seen a male covering eggs in the female's absence, suggested that males may occasionally incubate, and Schuster (1953) also recorded incubation as exceptionally being by the male. We have no evidence for this in Wales; all birds that we have lifted from clutches or young broods have been females.

The incubation period in Dippers in Britain is generally 16 or 17 days, with a range of 12–18 days (Shaw, 1978). Robson (1956) gave a mean of 17 days with no record of less than 16 days, and Mawby (1961) found a range of 15–18 days for 35 clutches, with a mean of 16.2. Studies elsewhere in Europe indicate an incubation period of similar length, e.g. a mean of 16.5 days (Zang, 1981) and 16.8 days (Schmid, 1985a), but incubation may be shorter later in the season. Thus in Switzerland, Breitenmoser-Wursten (1988) found a mean of 17.7 days for first and repeat clutches, but only 16.6 days for second clutches. She commented that the incubation period was longer at higher altitudes.

Incubation in the American Dipper is also usually 16 days, with a range of 14–18 days (Price and Bock, 1973); Steiger (1940) found incubation at one nest was only 13 days. Haneda and Koshikara (1969) recorded an incubation period of 17 days for Brown Dippers in Japan.

Where eggs are infertile, females may continue to sit long after the normal incubation period. Mawby (1961) recorded one bird sitting for 28 days before deserting her clutch, and Efteland (1975) for as long as 42 days.

The female may spend periods on the nest before the clutch is complete, possibly incubating from the penultimate egg; she also usually roosts in the nest during the laying period. This may explain the somewhat asynchronous hatching at some nests, nestlings hatching over a period of one to two days (e.g. Rankin and Rankin, 1940). Price and Bock (1973) also recorded an instance of asynchronous hatching in a brood of American Dippers.

Although the female leaves the nest for short periods during incubation, there are records of the male bringing food to the nest for her (Efteland, pers. comm.). The male, as noted above, spends much time 'on guard', perching close to the nest. Efteland found that males spent 40% of their time in such a way during incubation. Mate-guarding will reduce the chance of a neighbouring male attempting to remove eggs or young from the nest. This would cause the female to have to relay and so give the neighbouring male a chance of copulating with her.

DIPPER BROODS

Brood Size

Most broods hatched in late April and were of four chicks (Table 11). However, many were also of five and this was the commonest brood size in breeding attempts hatched up to 15 May. The mean brood size (at 7–14 days) for Dippers in south Wales was 4.13 (Table 12), larger than found in most other studies in Britain (Tyler and Ormerod, 1985), but similar to the brood size found by Hardey, Rae and Rae (1978) in Scotland. It was smaller though than that found in Norway, in Austria, parts of Germany and in the Swiss Alps (Efteland

Table 11. The brood size and hatching dates of Dippers in Wales

Brood size	*All sites*	*Early season*	*Late season*	*Acid streams*	*Circumneutral streams*
1	2.0	2.1	1.7	3.4	1.6
2	9.3	6.5	18.1	11.3	8.6
3	21.4	18.6	30.5	24.9	20.3
4	35.4	35.9	33.9	40.1	33.9
5	31.3	36.2	15.8	20.3	34.8
6	0.5	0.7	0	0	0.7
No of broods:	743	566	177	177	566

Time of hatching (period)	*All sites*	*Acid streams*	*Circumneutral streams*
<31 March (1)	2.8	0.6	3.5
1–15 April (2)	19.8	13.5	21.7
16–30 April (3)	32.2	22.6	35.1
1–15 May (4)	21.2	33.9	17.3
16–31 May (5)	13.4	16.9	12.4
1–15 June (6)	9.8	11.3	9.4
16–30 June (7)	0.8	1.1	0.5
Total *n*	743	177	566

NOTES:
Early and late season nests are taken before or after 15 May, the date at which the modal brood size shifted from 5 to 4. Acidic and circumneutral sites were divided at pH 6.5. (Identical sample sizes in the columns in these divisions are coincidental.) The values are all percentages of the total sample.

and Kyllingstad, 1984; Schmid, 1985a; Breitenmoser-Wursten, 1988; Sackl and Dick, 1988). Brood size of the nominate *cinclus* and of *aquaticus* in these countries may possibly differ from that of *gularis*, the 'race' occurring in most of Britain; but, as for clutch size, brood size is larger at higher latitudes, as in Norway.

Within years, broods from clutches started in May and June in our Welsh study area were significantly smaller than those from clutches started earlier, so following the seasonal change in clutch size. Other workers have found a similar seasonal effect on brood size. Zang (1981), for example, recorded a decrease of 0.13 young for every ten days of the season, although Sackl and Dick (1988) could demonstrate no such effect. Zang (1981) found broods were largest at 400–500 m, brood size decreasing above and below this altitude.

The quality of the territory affects brood sizes as it does clutch size. Thus a comparison of good quality (pH >6) and poor quality acidic streams (pH 4.5–6.0) in Wales in 1985 and 1986 showed brood sizes on the latter to be significantly smaller (mean of 3.6, SD 0.9, $n = 22$) than on the former (mean of 4.2, SD 0.8, $n = 132$); this difference was explained by the smaller clutch sizes at acidic sites

Table 12. Mean brood size of Dippers in 14 different studies

Region	*Mean*	*SD*	n	P	*Brood/clutch*	*Source*
SW Norway	4.3	NA	72	NA	85	Efteland and Kyllingstad (1984)
Grampian, Scotland	4.16	NA	64	NA	91	Hardey, Rae and Rae (1978)
Germany	4.15	–	–	–	83	Rockenbauch (1985)
SE Wales	4.13	1.04	286	–	86	Tyler and Ormerod (1985)
Swiss Alps	4.07	–	–	–	83	Breitenmoser-Wursten (1988)
Germany	4.03	–	–	–	85	Schmid (1985a)
E Germany	3.91	NA	66	NA	81	Zang (1981)
Midlothian, Scotland	3.81	0.83	27	0.05 (*U*)	NA	Serle and Bryson (1935)
Britain	3.55	1.21	637	0.001 (*t*)	80	Shaw (1978)
N Yorkshire/Cumbria, England	3.5	1.19	40	0.02 (*U*)	81	D. Baines (pers. comm.)
Lower Austria	3.38	1.97	42	—	75	Sackl and Dick (1988)
Cumbria, England	3.26	NA	40	NA	78	Robson (1956)
Grampian, Scotland	3	NA	15	NA	88	Hewson (1967)
Derbyshire, England	2.69	NA	26	NA	62	Shooter (1970)

NOTES:

SD = standard deviation; *n* = sample size; *P* indicates level of significance of the difference between each study and the present one, using *U*-test as indicated; brood/clutch = mean brood size as a percentage of mean clutch size.

(Ormerod and Tyler, 1987b; Ormerod *et al.*, 1991). Zang (1985) also found larger broods on the more productive rivers on south-facing slopes in the Harz Mountains in eastern Germany, than on rivers on north-facing slopes.

The mean brood size may differ slightly from year to year. Thus, in the south-east Wales study area the annual mean size of broods hatching from clutches laid before the end of April between 1975 and 1993 ranged from 3.9 to 4.6, with a combined mean from 408 broods of 4.28 (Fig. 44a). When the data from the 19 years were lumped into three periods (one of seven years and two each of six years), a small but significant ($P < 0.01$) decline was evident from 4.43 chicks per brood in the earliest period (1975–81) to 4.07 in the last six-year period. The reasons for this decline in brood size of less than half a chick, and its significance in terms of recruitment of young to the population, are unclear (see Chapter 8).

Adaptive Significance of Early Laying and of Brood Size

Brood size and timing of laying of any species of bird are finely tuned to the species' general ecology. Lack (1947; 1948a, b), hypothesised that birds have clutches and broods of a size that will maximise the number of offspring that can be raised, and suggested that the most productive brood size would also be the commonest in the population. He also believed that breeding should be timed to confer the greatest chances of survival on offspring, through being coincident with optimal conditions, i.e. the time of breeding is such that there will be a plentiful supply of food available during the nestling period and when juveniles are out of the nest (Lack, 1954). Various modifications have been put forward to explain an apparent anomaly found in some studies whereby the commonest brood size is smaller than the most productive. One such modification of Lack's views is the 'individual optimisation' hypothesis which suggests that individual birds phenotypically adjust their brood size (and presumably time of breeding) to suit conditions on their territory (Hogstedt, 1980).

To test Lack's hypotheses and modifications, we looked at the post-fledging survival of Dippers from broods of different sizes and from different periods of the breeding season, and assessed the relative productivity of different brood sizes from different times of the breeding season. We also assessed whether adjustments to the breeding strategy along poor quality, acidic streams and good quality, circumneutral streams are adaptive in promoting survival of the offspring. We argued that if optimisation were occurring to suit these conditions of contrasting territory quality, then the most productive broods at acidic sites would be smaller than at circumneutral sites, and that advanced and delayed laying along acidic and circumneutral streams, respectively, would enhance post-fledging survival.

From ringing numerous nestling Dippers and subsequently recapturing them as independent young or adults, we found evidence that post-fledging survival was higher among broods in the peak period of hatching (16–30 April) than from

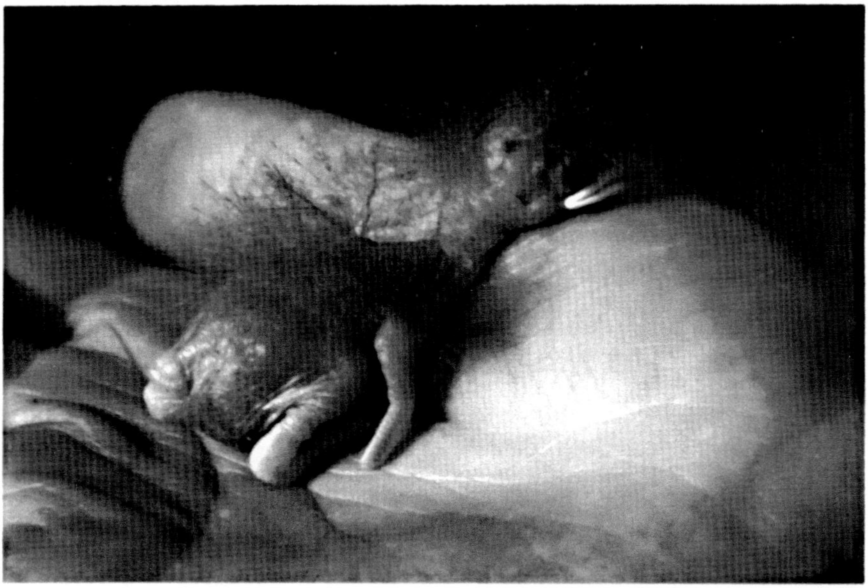

Dipper nestling at one to two days

earlier or later periods of the season (Fig. 46) (Ormerod and Tyler, 1993c). This indicates that the pattern of laying in the Dipper is such that it confers the greatest chances of survival on young birds, and is thus consistent with Lack's (1954) hypothesis.

When we looked at the survival of 2871 ringed nestlings from a range of brood sizes, we found that in general the proportion of young recaptured as independent young or as adults declined with increasing brood size, the exceptions being broods of 1, from which no recoveries occurred, and broods of 6 (Fig. 47). Omitting these broods whose sample sizes were very small, and allowing for a seasonal correction, broods of 4, the most frequent brood size overall, proved to be the most productive. However, there is a seasonal change in brood size, with broods of 5 being the most common brood size before mid May and broods of 4 after this date. When the productivity of different brood sizes was compared before and after mid-May, broods of 5 were found to be the most productive in the earlier period and broods of 4 after 15 May (Fig. 48). The proportions of recaptures from different brood sizes also showed differing patterns at acidic and circumneutral sites. The most productive brood sizes at the former poor quality sites were 4 followed by 3, whereas at circumneutral sites brood size 4 followed by 5 was the most productive (Fig. 49). Birds at both types of sites thus tended to have broods at or near the size that would be most productive. Dippers at acidic sites laid significantly later than at circumneutral sites, but post-fledging survival declined most rapidly through the season at the former.

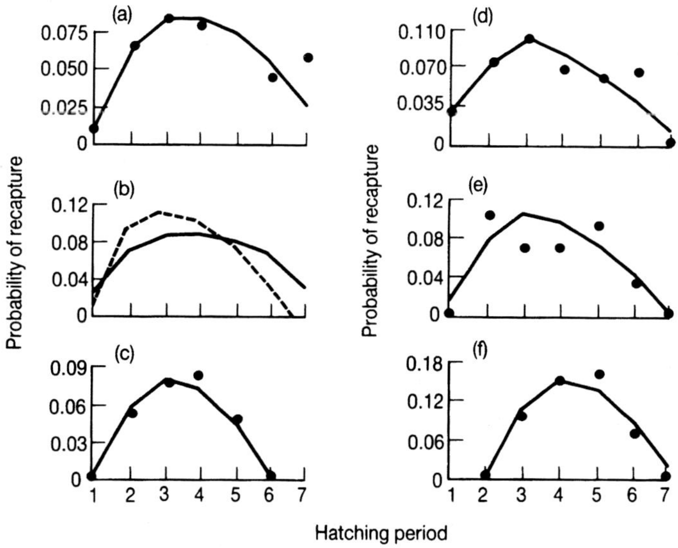

Figure 46. The probability of recapturing Dippers after fledging in relation to their times of hatching in upland Wales.

NOTES: Closed circles show real data points, while the fitted curves are based on the quadratic regressions listed in Appendix 2. (a) All broods; (b) Acidic ---- and circumneutral —— streams; (c) broods of 5; (d) broods of 4; (e) broods of 3; (f) broods of 2. Times of hatching are shown as the seven periods listed in Table 11 (Ormerod and Tyler, 1993c).

In general, therefore, our survival data show that patterns of brood size in the Dipper support Lack's hypotheses. They provide evidence from both seasonal and spatial patterns that brood size in the Dipper may be optimised in ways consistent with the enhancement of productivity. By contrast, delayed breeding at acidic sites contrasts with the pattern expected from optimisation, instead reflecting resource scarcity.

Brooding Behaviour

At hatching, the young are naked and helpless (altricial), and are brooded by the female alone. For the first few days, the female only leaves the nest for a few minutes at a time after brooding periods of 20–65 minutes, occasionally up to 113 minutes (Rankin and Rankin, 1940). Eggbrecht (1937) found that during one day's incubation, soon after the young had hatched, the female only left the nest 20 times, at average intervals of 32 minutes, and was off the nest for about 9 minutes at a time. Brooding periods decrease as the nestlings grow, generally ceasing by the time they are 12–13 days old.

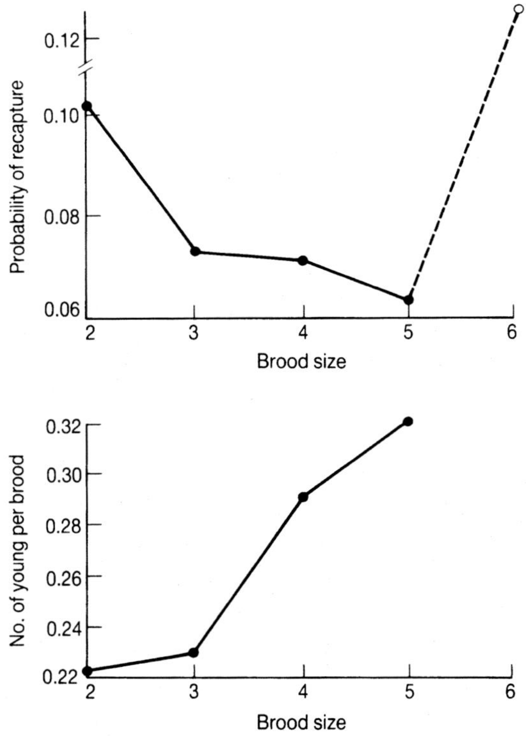

Figure 47. (a) The probabilities of recapturing Dippers after fledging from broods of different sizes, and (b) an index of the resulting number of young per brood (probability of recapture × brood size).

NOTES: The open symbol from broods of six reflects a small sample. These values are not corrected to account for seasonal changes in brood size and survival (Ormerod and Tyler, 1993c).

FEEDING OF THE NESTLINGS

Both Dipper parents feed the young, although during the first few days when the female is brooding, it is the male that makes most visits with food. He may feed the nestlings directly or pass food to the female. Dick and Sackl (1988) noted that at later stages of nestling growth, females paid longer visits to the nest than did the males, but visit lengths declined with nestling age.

Rankin and Rankin (1940) found that, on average, the young were fed 6 times per hour, with no real increase in frequency as the young grew. Rates varied from just 2 times an hour for very small nestlings to as much as 18 times per hour by the female for well-grown nestlings.

Our data on brood provisioning show some evidence of an increase in feeding

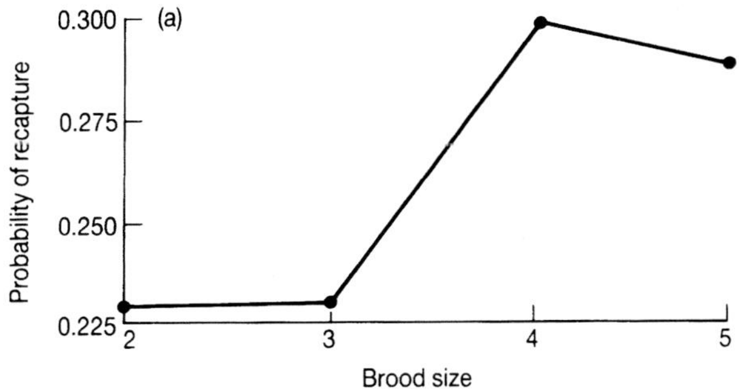

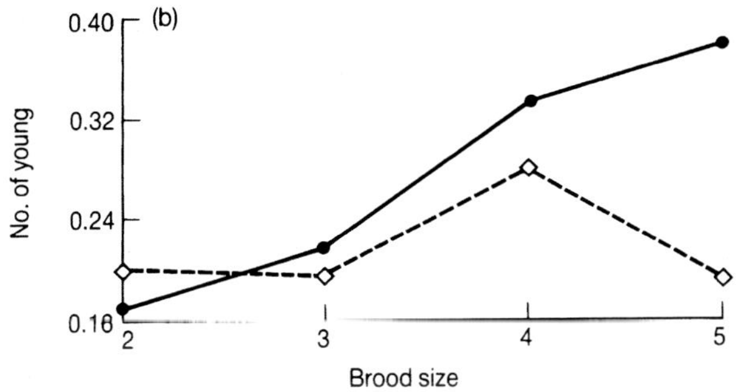

Figure 48. Indices of productivity of different brood sizes of Dippers (a) adjusted to take account of seasonal changes in the probability of recapture, and (b) broods hatching before (●) and after (◇) 15 May (Ormerod and Tyler, 1993c).

frequency when chicks were older than eight days, around which time brooding becomes very infrequent (see Chapter 4); as expected, the effect was more pronounced in females than in males (O'Halloran *et al.*, 1990). At sites where both sexes of the pair were feeding the young, the proportion of feeding visits made by the male did not differ significantly from the proportion made by females either early or late in the nestling period. Nor did the mean visit rate per hour differ between the sexes early in the brood period, although sex differences approached significance later in this period. Female feeding frequency was, not surprisingly, greater where no male was helping with provisioning, for example when she was widowed or when her mate was polygynous.

Kovshar (1979) found that male and female Brown Dippers visited the nest equally; he also noted an increase in the frequency of feeding as the young grew. Older young were fed almost 40 times per hour throughout the day.

Many factors, for example brood size and age, the amount and quality of food available and the time of the season, may affect feeding frequencies in passerines. Smaller and younger broods generally require less food than larger ones; a

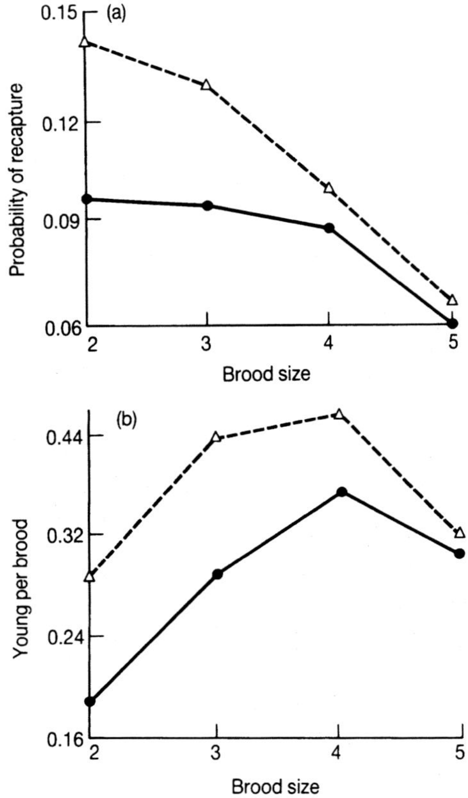

Figure 49. (a) The probabilities of recapturing Dippers after fledging from broods of different sizes, and (b) an index of the resulting number of young per brood along acidic (△) and circumneutral (●) streams.

NOTES: These values are corrected to take account of seasonal changes in brood size and survival. Note that only the relative effects of brood size are comparable between acidic and circumneutral streams (Ormerod and Tyler, 1993c).

plentiful supply of large prey items, such as some caddis larvae, would mean fewer visits by Dippers than at sites where the adults were collecting small stoneflies or blackfly larvae. Later in the season, longer hours of daylight during which adults can feed young may reduce feeding frequencies.

Feeding sequence of the Nestlings and Nest Sanitation

When the young are small, adults enter the nest with food, approaching with a rattling call. The chick that is fed retreats to the rear of the nest, allowing another

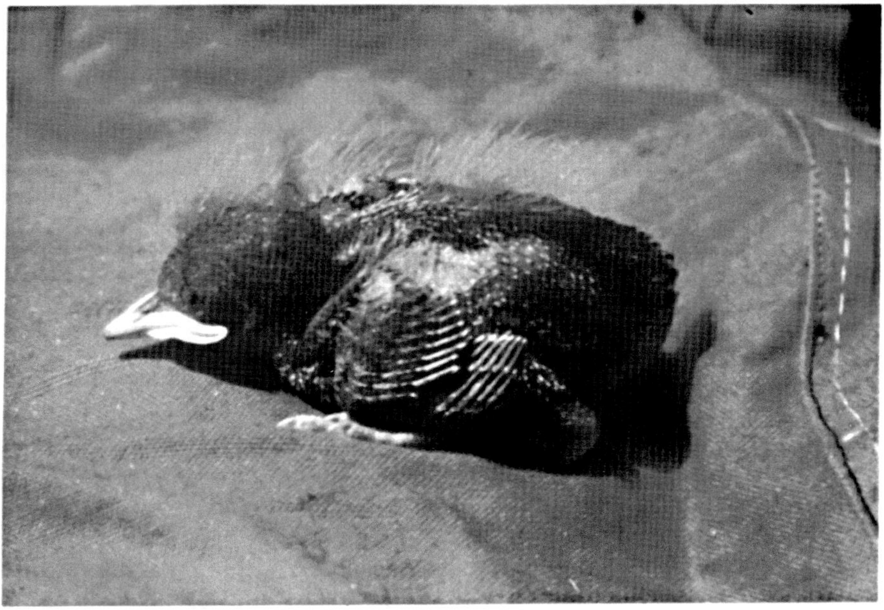

Dipper nestling at about 8 days

to be near the nest entrance (Sokolowski, 1964), a carousel system of feeding found also in the Kingfisher. The adult removes faecal sacs, dropping these into the water as it flies away. Dick and Sackl (1988) found that females were more concerned about nest sanitation than males, but recorded one female accidentally removing a nestling with a dropping. We have recorded similar incidents in Wales where young birds were thrown out during nest-tidying. Older nestlings, as well as being heavier and more difficult for the adults to eject, have a strong grip which makes it difficult to remove them from the nest. They cling tightly on to the nest-lining or on to their siblings.

Deliberate infanticide was recorded by Sonya Yoerg, who watched a male Dipper from an adjacent territory fly to its neighbour's nest and remove a nestling. The removal of the whole brood in this way could lead, as mentioned above, to the female relaying, and hence give the neighbouring male the chance to mate with her and father the resulting brood. Wilson (1992) also recorded a probable case of sexually-selected infanticide in Dippers.

From about seven days, the young move to the entrance of the nest when they see a parent approaching; they call loudly and beg for food, revealing yellow-orange gapes. The conspicuous white flanges to their bills also emphasise the gaping mouths. The food calls given by the nestlings can be very loud and persistent, and have been described as shrill and squeaking by Shaw (1978). The visiting adult clings briefly to the nest while it feeds them, and for a few days more continues to take away faecal sacs from the nest. The young at 9–10 days turn and defaecate out of the nest, the sacs usually falling into water to be swept downstream or to sink to the stream bed. If there are exposed rocks or if the river level has fallen since nest building, attention can be drawn to the nest by the pile of white droppings on the ground, or smeared on to the cliff or wall below the nest. For a time, the adults will pick up fallen faecal sacs and carry these away, but towards the end of the fledging period they cease to do this. It was of interest, therefore, to observe that a pair of Rufous-throated Dippers feeding well-grown nestlings (*c.* 18–20 days), still collected and carried away faecal sacs.

THE FLEDGING PERIOD

Dippers, in common with other species which either nest in holes or have a domed nest, have a longer development period as nestlings than open-nesting passerines. Thus Dipper nestlings remain in the nest for 20–24 days, on average 22 days (Shaw, 1978) or 23 days (Robson, 1956). Schmid (1985a) recorded fledging periods of 23–27 days in Germany, and Breitenmoser-Wursten (1988) a mean of 21.5 days in the Swiss Alps. Baake (1982) found that the fledging period was dependent on the size of both the nest and the brood, being longer where nest and brood were larger. Fledging periods for the American Dipper appear to be slightly longer, with periods of 24 and 25 days noted by Bakus (1959a) and of 23–28 days (mean 24.6) for 10 broods by Price and Bock (1973). Asami and Haga (1983) recorded a fledging period of 21 days in the Brown Dipper.

Young Dippers can leave the nest and survive after only two weeks. If disturbed by a predator or human observer approaching too close, well-feathered nestlings 'explode' from the nest, jumping out into the water. These young birds can swim and dive superbly even though unable to fly at this age. They may swim or be carried for some distance downriver, but generally manage to scramble to the edge of the river where they hide below rocks or in vegetation. Their calls attract their parents which continue to feed them. Although more vulnerable to predators on the ground, some undoubtedly survive as shown by a 15-day-old nestling which was ringed and then 'exploded' from the nest, to be subsequently caught again as a flying juvenile by Balat (1964). Other young birds may scramble out on to ledges or into holes near the nest, only to return later. Even after fledging, some young return to the nest to roost.

Dipper nestlings gain weight rapidly, *c.* 3–4 g per day, until Day 13/14, after which their growth rate levels off (Figs. 50a and 51). Tarsal length also increases rapidly until Day 9/10 when the tarsus is fully-grown (Fig. 50b). Patterns of weight gain of Welsh and Irish Dippers are almost identical, although the rate of weight gain of nestlings in Wales appears to be higher than that of nestlings in Ireland (O'Halloran, Smiddy and O'Mahony, 1993). There is a significant effect of date on both the daily increase in weight and in daily tarsal growth, with

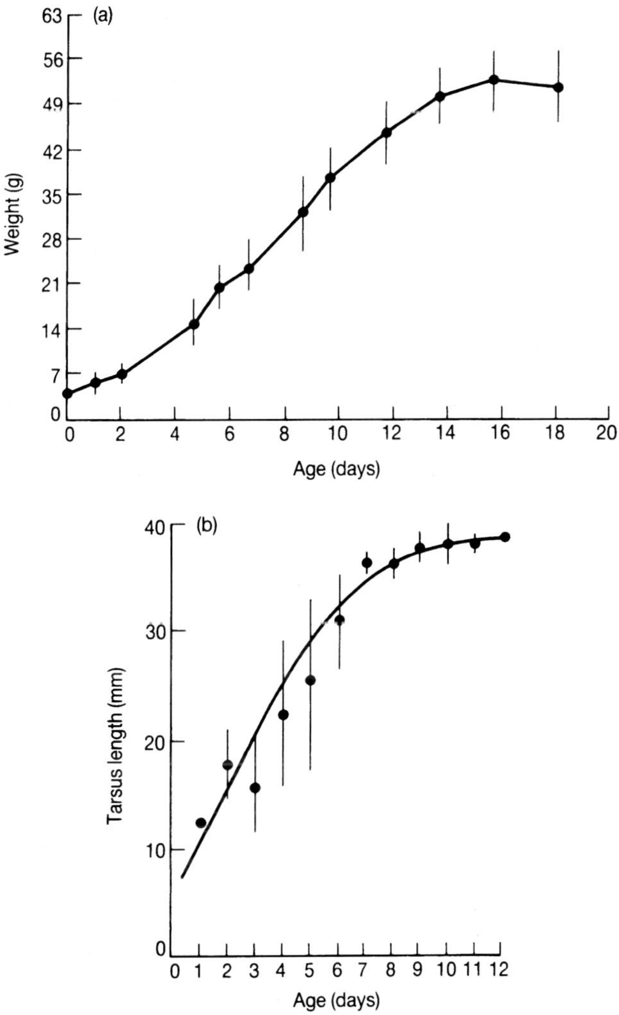

Figure 50. (a) The mean weights (±1 SD, n = *6 to 75), and (b) tarsal lengths of Dipper nestlings whose ages were accurately known or were assessed from nestling development (Ormerod, Tyler and Lewis, 1986). (Mathematical expression for these curves are available in Ormerod and Tyler, 1991a.)*

nestlings from early clutches (i.e. those laid in March) having a higher daily growth rate than nestlings from later clutches. At acidic sites there is also a reduced rate of growth of nestlings by comparison with that on circumneutral streams (Ormerod and Tyler, 1987b).

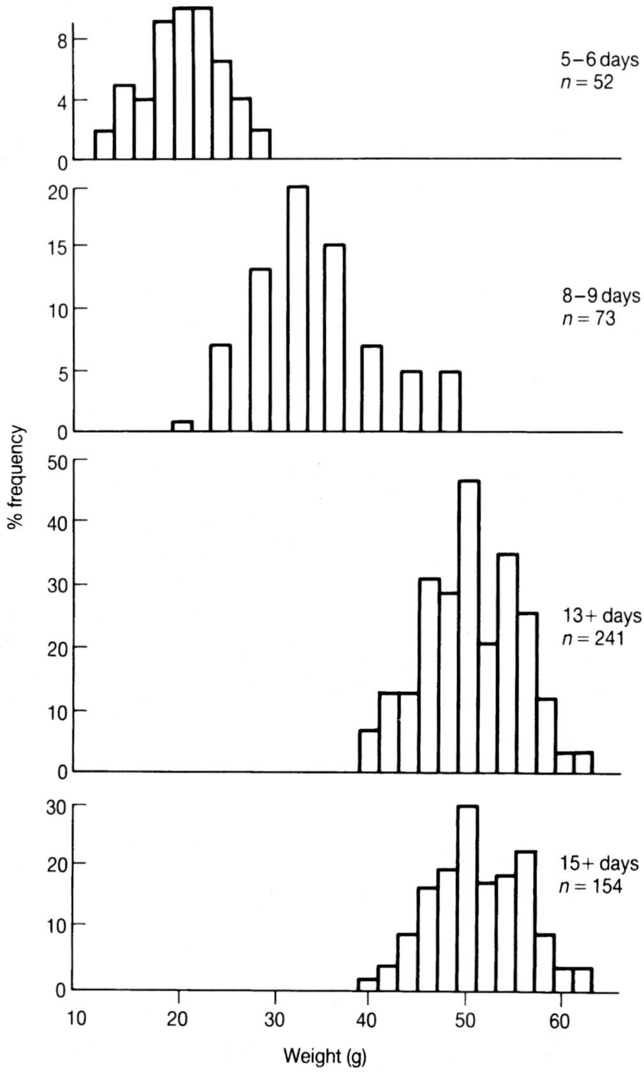

Figure 51. Frequency distribution of the weights of Dipper nestlings at selected ages (Ormerod, Tyler and Lewis, 1986).

BREEDING SUCCESS

Hatching and fledging success of Dippers are both comparatively high for a passerine. This can be largely explained by the domed nest which reduces predation compared with open cup-nesting species such as the Grey Wagtail. Hatching success of Dipper eggs varies from 61% (Shaw 1978) to 75 or 76% (Robson, 1956; Breitenmoser-Wursten, 1988). Where total failures are excluded, it may be as high as 87.5% (Shaw, 1978). Fledging success was given by Shaw as 84% (all nests) and 95% (excluding total failures), similar to 81% noted by Breitenmoser-Wursten (1988) and 94% by Robson (1956). Figures for overall breeding success, defined as the percentage of nests that produced at least one fledged young, range from 56% (Efteland and Kyllingstad, 1984; Breitenmoser-Wursten, 1988) to 68.4% (Schmid, 1985a). Where success is calculated as the percentage of all eggs producing fledged young, this varies between studies from 51 to 75% (Robson, 1956; Shooter, 1970; Shaw 1978; Efteland and Kyllingstad, 1984).

Schmid (1985a) and Breitenmoser-Wursten (1988) give the mean number of young produced per nesting attempt as 3.23 and 3.69, respectively, but Mawby (1961) recorded only 2.7 for first broods and 2.0 for second broods. The mean number per successful nest was noted as 2.93 and 4.03 by Sackl and Dick (1988) and by Schmid (1985a).

Many authors comment on the variation in breeding success from year to year, and the greater success of first broods by comparison with replacement or second clutches. In Wales first broods are much more successful than second broods; we found more desertions among later broods, higher mortality among nestlings of later broods and a higher predation rate. Breitenmoser-Wursten (1988) however, found the reverse. She found nests were more likely to be successful as the season progressed, with a 6.4% increase in success per 10 days. Success was influenced greatly by stream structure (safety from predators and flooding), with from 8 to 40% of first nests being lost.

Sackl and Dick (1988) noted that survival of nests was much lower during the second half of the nesting period (36.1%) than during incubation and first days of the nestling period (81.9%). They attributed this to the older broods being noisier and hence attracting the attention of predators.

Schmid and Spitznagel (1985) found that pairs of south German Dippers with accentuated size differences showed significantly higher reproductive success than pairs where males and females differed only slightly in size and weight.

INCIDENCE OF SECOND BROODS

In Wales we found on the lower Wye tributaries that 19.3% of nests were second clutches or broods (Tyler and Ormerod, 1985), but birds at higher altitudes and/or on poor quality sites, which started laying later, rarely had a second brood. On average, 24% of birds at lower altitudes or on good quality sites had second broods (Ormerod and Tyler, 1987b), but in some years on favoured streams 80–90% of pairs attempt a second brood. Values for other

Table 13. Incidence of second broods (percentage of pairs having second broods)

Area/county	*Mean*	*Number of nests*	*Source*
UK			
Westmoreland	16.6		Robson (1956)
Wales	19.3	403	Tyler and Ormerod (1985)
	0–85		Tyler and Ormerod (unpubl.)
NW Yorkshire	20		Mawby (1961)
Scotland (Grampian)	32.8	58	Hardy, Rae and Rae (1978)
Cumbria	53.4		D. Baines (pers. comm.)
Scotland	60		Da Prato and Langslow (1976)
Ireland	<5		O'Halloran, Smiddy and O'Mahony (1993)
EUROPE			
Sweden (1973)	5.3	38	Mork (1975)
Sweden (1974)	6.5	62	Mork (1975)
Czechoslovakia	10.8		Balat (1964)
Germany	11.8		Zang (1981)
Austria	15.9	63	Sackl and Dick (1988)
Swiss Alps	18.1 (3.3–37.5)	150	Breitenmoser-Wursten (1988)
Poland	33	30	Czapulak (1988)
Germany	49		Schmid (1985a)

studies in Britain range from less than 5% to 60%. In Sweden as few as 5.3% of pairs attempted second broods (Table 13). The high incidence in Cumbria was during a year when a large proportion of first clutches was started in March (D. Baines, pers. comm.). Da Prato and Langslow (1976) related the higher incidence (60%) of second broods on lower reaches of the River South Esk in Scotland, compared with upriver reaches, to the better availability and composition of the food supply at lower altitudes.

As in our study area, Wagner (1985) found that Dippers in Austria which started laying late in the season, and those nesting above 500 m, rarely had a second brood. The low incidence of second broods in Sweden (Mork, 1975) is probably because of the late start to, and short duration of, the breeding season at high latitudes. Breitenmoser-Wursten (1988) also found that second clutches were only possible after an early first brood and in favourable weather conditions. They were limited in the Swiss Alps to birds breeding below 1300 m and varied greatly between years (range 3.3–37.5%). Schmid (1985a) found more second broods at 350–400 m a.s.l. than above or below this altitude.

The Brown Dipper in Japan has only one brood a year (Asami and Haga, 1983), but more data are needed from other parts of its range. The American Dipper is frequently double-brooded, although Hann (1950) and Bakus (1959a) both believed that only one brood was raised ordinarily. Drew (1881) first recorded double-broods in this species, whilst more recently Price and Bock (1983) found during a four-year study that about 40% of Dippers in the Front Range in Colorado had second broods, although the proportion of pairs attempting a second brood varied between years (5.5% to almost 70%). Their figures, however, included pairs which relaid following a failed first nest as well as genuine second broods, so the incidence of double-broods is less than that given above.

Pairs of Dippers in Britain or Europe that started first broods in February or early March could conceivably just fit in three broods before they undergo moult in late July or August. We know of no instance though from our large sample of pairs in Wales where three broods have successfully fledged. Pairs not infrequently lay three clutches but invariably only when one has failed. There have been records of three broods, such as in Moorhouse Reserve by Maurice Waterhouse, but we believe that this was due to failure of an earlier nest rather than three true broods. We doubt whether three broods are ever reared by one pair of Dippers in Britain; if so, it must be extremely rare. Fuchs (1970), however, recorded a pair rearing three broods successfully within one season in Germany, and Grunwald (1973) also noted a pair which was triple-brooded in the same nest-box.

Recently fledged juvenile Dipper

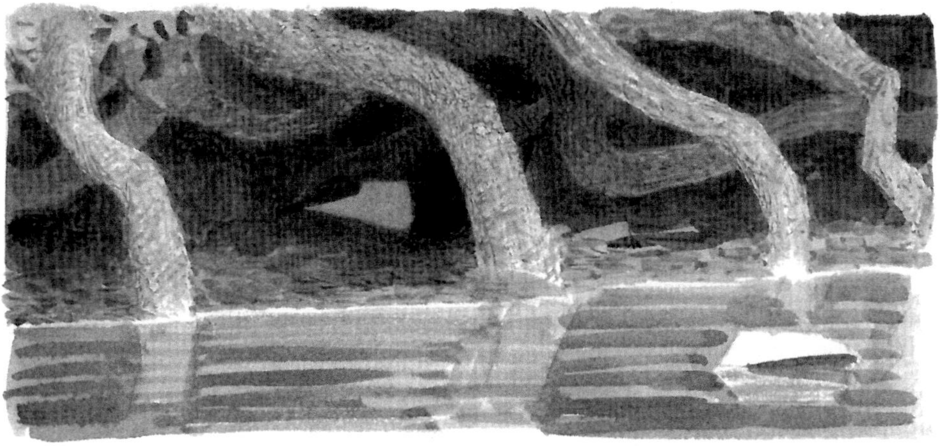

Moulting Dipper skulking below tree roots

CHAPTER 7

The post-breeding phase

This chapter covers the development and behaviour of juveniles, their moult, dispersal patterns and mortality. We also look at the rapid way that adults moult, replacing their old worn feathers immediately after the breeding season, and the survival strategies of dippers for the winter months. These strategies include hard weather or altitudinal movements as well as true migration, with some birds regularly flying distances of up to 1000 km.

JUVENILES

Juvenile (White-throated) Dippers differ markedly in plumage from the adults. The head and mantle of a juvenile is grey-brown and its underside is buffish with dark spots and smudges, due to brownish grey tips to the breast feathers. When the juveniles first leave the nest, wispy down is apparent particularly on their heads. Young Brown Dippers show buff tips to their coverts, but American Dipper juveniles closely resemble the adults apart from being paler below and from a difference in bill colour; the bills of the young birds are at first pink, becoming darker with age. The juvenile plumage of Rufous-

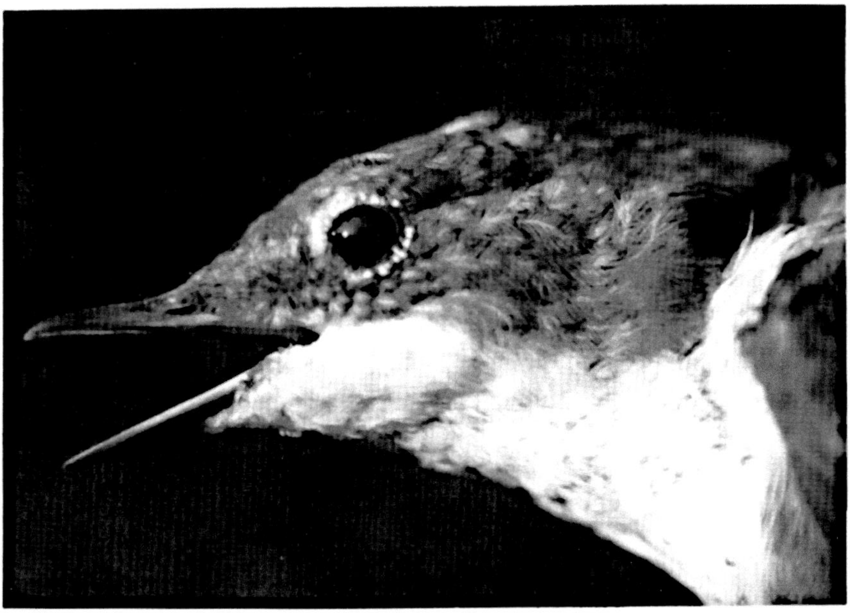

Close up of head of juvenile Dipper

throated Dippers was not described (Fjeldsa and Krabbe, 1990) until our visit to Argentina in 1993. The plumage of two well-grown nestlings proved to be very similar to that of the adults, with the rufous throat already apparent.

The Period of Dependence of Juveniles on Their Parents

The newly fledged young Dippers are dependent on their parents for food for only one to two weeks. Robson (1956) recorded families staying together for a week after fledging and Fuchs (1970) also found most young were independent about a week after leaving the nest; he did not see them being fed by their parents after 11 days. Balat (1964) and Andersson and Wester (1975), however, noted that young birds were not independent until 11–18 days after fledging. In Wales many juveniles appear to be at least partly dependent on one or both parents for up to two weeks.

Before independence, the fledged brood may be split between the two parents, with one adult feeding one to three juveniles and the other parent some distance away feeding the rest of the brood. Other observers, such as Fuchs (1970) and Sudhaus (1972) found that the female fed the juveniles more than did the male.

Juveniles whose parents then have a second brood, may receive less attention than young of single-brooded parents or from second broods. We have found that there is usually a 1–2 week period between the fledging of one brood and the laying of the first egg of the next clutch. In this period the female relines the old nest or, more rarely, the pair may build a new nest. As the female is thus occupied soon after her first young have fledged, the male may take much of the responsibility for feeding the juveniles. Kovshar (1979) also remarked that in white-bellied Dippers, fledged broods may be attended by only one parent.

Juvenile Behaviour

For the first few days, the juveniles spend much time quietly sitting on rocks or hiding in crevices at the river's edge, resting and preening within 100 m of the nest, whilst the parents collect food. When a parent approaches, they move towards it, calling repeatedly and begging for food; they also quiver or flutter their wings, in a similar way to the wing-fluttering of females during courtship. As they grow, they more frequently follow their parents, continuously calling and begging for food, wing-fluttering and bobbing. They occasionally and inexpertly peck at food items or other objects on rocks or the ground, but such foraging bouts are short. Even when food passed by a parent is dropped by a juvenile, it does not pick up the item, merely waiting for its parent to do so. Sonya Yoerg found that juveniles, when about five days out of the nest, spend much more time feeding than younger birds, the bouts of foraging becoming longer with increasing age (see Chapter 4). Blackflies or simuliids, which can occur at very high densities on rocks in the river, are an important food item during this period of learning (see Chapter 4). Adults may still continue to provision their young until about two weeks post-fledging, and by three weeks the juveniles are able to use a variety of foraging techniques and survive on their own.

Initially, the young birds dive into the water or hide among rocks if danger threatens, but by the time they are a week old they can fly more strongly and escape in this way. Kovshar (1979) recorded a brood of white-bellied Dippers remaining within 100 m of the nest for nine days, and then the female escorted the juveniles, one by one, to a lake 400 m away. She flew 20–30 m ahead of each juvenile, pausing repeatedly to wait for it to catch up.

Juvenile Moult

Once the young are able to fend for themselves, some still remain within their parents' territory until after they have completed their juvenile moult at about 4–6 weeks after fledging. Some continue to return to their old nest or close to it to roost several weeks or months after fledging. During the juvenile moult, the body feathers and lesser, median and inner greater coverts are replaced but flight feathers and primary coverts are not. The young bird after moulting resembles the adult but can be distinguished from older birds throughout its first winter by the white 'juvenile' fringes on the retained primary and outer greater coverts. These wear gradually so that by the spring many first-year birds are indistinguishable from adults; others still show a pale edge to the coverts at this time. As in Dunnocks, the eye colour of young Dippers is a more olive-brown compared with the chestnut-brown of the adult eye. Some authors (e.g. Rockenbauch, 1985b) claim that eye colour continues to redden as adults age, so that individuals can be accurately aged on this character. Rockenbauch recognised six eye colours from olive-brown to brown to dark brown to red-brown.

Growth of Juveniles

Young Dippers, although fledging at adult weight, have shorter wings than adults. Growth of their wings continues for several months after they have left the nest, with young birds, particularly the males, still having significantly shorter wings than adults in the late autumn (Ormerod, Tyler and Lewis, 1986). Other passerines which do not have a post-juvenile primary moult also have shorter wings than adults and this has been explained by some authors as due either to growth of first-year birds being constrained by their foraging ability or by undernourishment in the nest, or alternatively by the inexperienced first-year birds somehow benefiting from the increased manoeuvrability provided by shorter wings. We do not know which, if either, of these explanations applies to Dippers.

Schmid and Spitznagel (1985) showed that, in addition to wing length, the length of the body, head, tarsus and bill, and depth of bill, all increase as the young birds grow. Only the footspan decreases as the claws become worn with age.

Few data are available for the American species or for Brown Dippers, but several young Brown Dippers that we caught in Nepal in November 1991 had significantly shorter wings than the adults we handled (Tyler and Ormerod, 1993).

Mortality of Juvenile Dippers

A high mortality of juveniles occurs in the immediate post-fledging period (Fig. 52) through predation and accidents (see Chapter 9), or possibly starvation (Galbraith and Tyler, 1982), with some 65% dying before their first winter. Serle and Bryson (1935) also commented on the high mortality in the post-fledging period, compared with 'the negligible losses in the egg and nestling stages'. Many other observers have noted the very low recapture rates of birds ringed as nestlings, attributing this mainly to high mortality, but also to dispersal of juveniles away from the natal area. Moss (1975) recaptured only 7% of his ringed nestlings, and Hewson (1969) only 8%. This figure is similar to our recapture

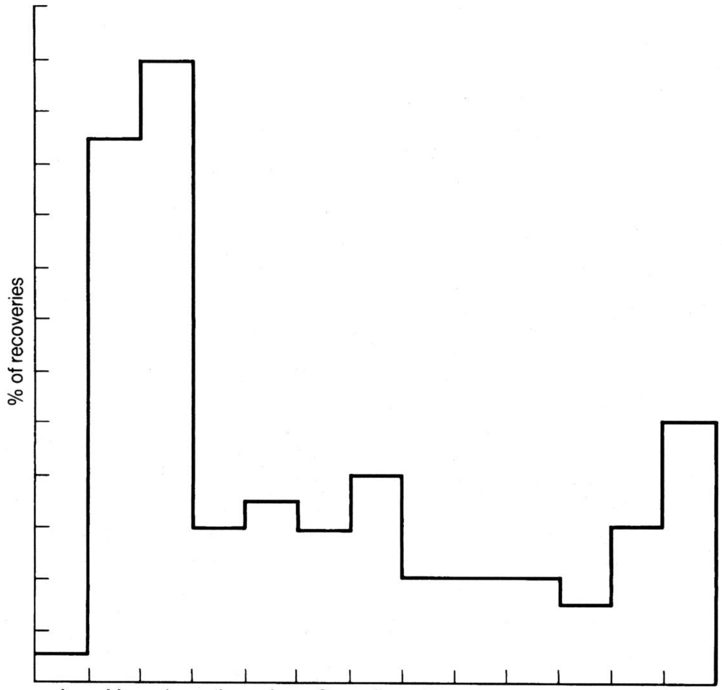

Figure 52. Seasonal changes in mortality of juveniles as shown by the timing of recoveries of birds ringed in their first year and dying before April of the year after hatching (Galbraith and Tyler, 1982).

rate of birds initially ringed as nestlings, but it does not necessarily equate to the rate of mortality. Loss of young to the study area through dispersal is, as mentioned above, an important factor too. We estimate that around 16–20% survival of young birds is necessary to keep the population stable.

Price and Bock (1983) similarly recaught a very small proportion of juvenile American Dippers, again attributing this to high post-fledging mortality, and then to dispersal. Of those young birds that had survived to the first year, 80% had left their natal area.

Dispersal of Juvenile Dippers

Juveniles leave their parents' territory either once they are independent at around two weeks or after the juvenile moult. Zink (1981) recorded most juveniles leaving at 6–7 weeks.

Although there are records of adults behaving aggressively towards juveniles, most breeding pairs will tolerate young birds originating from other pairs, within their territories. Balat (1962b) noted that juveniles could cross territorial boundaries without being attacked by the territory-holders. Jeremy Wilson (pers. comm.) also found a low incidence of aggression both between juveniles and between juveniles and adults during natal dispersal. Although there is little aggression between dispersing young birds and adults, intrasexual competition among males may still occur through, for example, song or passive avoidance of territories already occupied.

The young birds may wander up or downstream. We found no significant difference in the likelihood of birds moving up or down (Table 14) although some observers claim that there is a downstream preference, and others that juveniles tend to move upstream (e.g. Nicoll, 1971). Schmid (1985b) showed a tendency for birds to fly downriver, and then up tributaries, sometimes crossing watersheds.

We have records of young Dippers fledging during early April, caught again at least 10 km away by late May. The major period for dispersal, however, appears to be during the late summer (Fig. 53) when the Welsh juveniles move up to 30 or 40 km. In Britain and in central and western Europe dispersing juveniles and young Dippers (races *gularis* and *aquaticus*) rarely move more than 40–50 km (Zink, 1981; Galbraith and Tyler, 1982). Up to 1991, the longest recorded movement of a British Dipper (*gularis*) from over 440 recoveries reported to the BTO, was only 58 km. This was a bird ringed as a nestling near Craven Arms in Shropshire and found dead near Llangollen, Clwyd in north Wales (Mead and Clark, 1989). Two other Dippers apparently moving over 50 km, one even initially reported as moving 116 km, have subsequently been found to have travelled part of that distance impacted on a car radiator! (see Mead and Clark, 1991; 1993). There has been a genuine movement of a Dipper within Britain further than the record 58 km, but this was made by a black-bellied bird of the nominate race, presumably a Scandinavian migrant, which moved 105 km from Burnham Market in Norfolk to Belstead in Suffolk (Mead and Clark, 1993). A migrant ringed in Sweden was caught in Fife in Scotland.

In Wales we have found marked differences in dispersal patterns between the sexes, with young males rarely moving more than 10 km from their natal site, and

Table 14. Movements of young male and female Dippers in relation to their natal site

Movements	*Male*	*Female*	*Total*
Up river	40 (48.70)	19 (33.93)	59 (43.58)
Down river	28 (34.15)	16 (28.57)	44 (31.88)
Down and then up	7 (8.54)	9 (16.07)	16 (11.59)
Across watershed	7 (8.54)	12 (21.43)	19 (14.76)
Totals	82	56	138

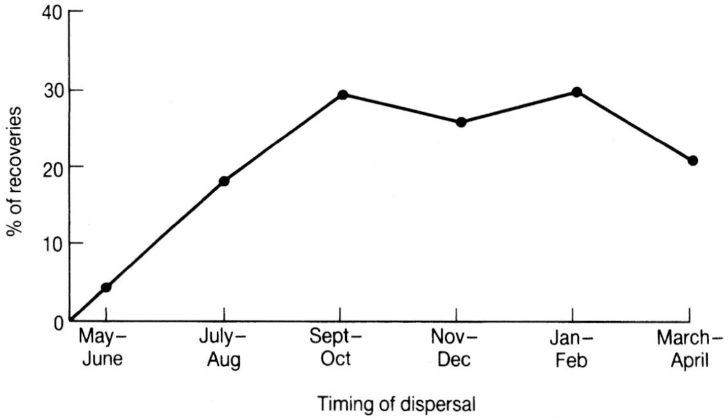

Figure 53. Timing of dispersal of Dippers in their first year: percentage of the recoveries at distances greater than 10 km in each two-month period (Galbraith and Tyler, 1982).

over 70% remaining within 6 km, but with young females often moving longer distances (Fig. 54) and not infrequently crossing watersheds (Tyler, Ormerod and Lewis, 1990). One of our broods produced three long-distance movements, all made by young females; these were recaught on different rivers 23–44 km from their nest on the upper R. Monnow, a lower Wye tributary, one flying south-east to Blakeney on the Severn Estuary, the second south-west on to the Usk system and the third north-west to a tributary of the middle Wye near Erwood. These movements are of course measured 'as the crow flies'. Dippers inevitably usually fly much further than these straight-line distances; typically they may follow many river meanders or move down a tributary, up or down the main river and then up another tributary.

The difference that we noted between the sexes in their pattern of natal dispersal remained consistent between the years 1984 and 1990 (Table 15). We did find some evidence though that males dispersed on average 30% further (*c.* 1 km) in years when populations were higher. No similar pattern was detected in females.

We showed too (Tyler, Ormerod and Lewis, 1990) that the observed distances for natal dispersal were significantly greater than those expected if the juveniles were simply moving far enough to find territories made vacant through adult mortality (Fig. 55).

JUVENILES OF OTHER SPECIES OF DIPPER

Juvenile American Dippers behave in a very similar manner to those of White-fronted Dippers, becoming independent by the fourteenth day, and moulting

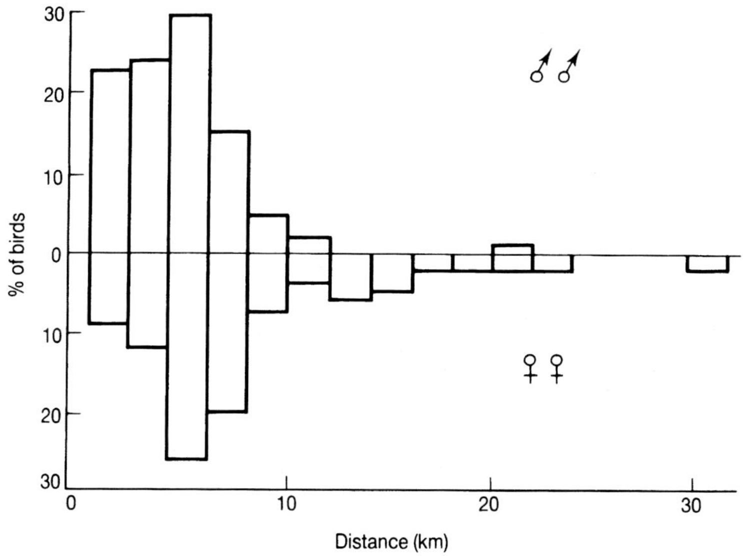

Figure 54. Natal dispersal in Welsh Dippers, showing percentage frequency distribution of movements of male and female birds initially ringed as nestlings (Tyler, Ormerod and Lewis, 1990).

about 6 weeks after fledging (Bakus, 1959b). As the European birds, they are paler grey than the adults with a lighter underside, but do not have the dark markings on the breast. Dispersal occurs once the young are independent, with Price and Bock (1983) finding that most moved upstream. Of 147 nestlings ringed, they recaught 42 (29%) within 3 months of fledging: 7 (17%) had moved less than 1 km, 10 (24%) had moved downstream and 14 (33%) upstream, but a further 11 (26%) had changed drainages, which Price and Bock took as an indication of their probably moving upstream first. They suggested the young birds moved upstream to better foraging areas. Some juveniles they found close to the nest 29 days after fledging; others moved rapidly away, such as one bird which had moved 8 km only 22 days after fledging. The distance that a young bird moved between its natal site and breeding site ranged from 2 to 74 km, with 10 km being the median distance and 17.8 km the mean.

The limited data available for other species suggests an almost identical pattern of post-fledging behaviour. Lasso (1988) wrote that juvenile White-capped Dippers only a few hours after fledging, follow their parents, begging for food. When begging, they crouched slightly, raised their wings to a horizontal level, erected their crown feathers, stretched their neck and opened their bill, screeching shrilly as they did in the nest when soliciting food. Juveniles preened frequently for 3–4 minutes at a time and for a total time of about one hour per day.

Table 15. Year-to-year patterns in the natal dispersal of Dippers in upland Wales

Year	*Males*	*Females*	F	P
1985	4.1 (0.19, 18)	8.5 (0.46, 21)	7.21	0.02
1986	4.1 (0.29, 29)	8.3 (0.35, 24)	11.25	0.002
1987	3.7 (0.31, 29)	8.9 (0.39, 24)	15.77	0.001
1988	2.8 (0.22, 19)	8.1 (0.40, 19)	19.19	0.001
1989	3.4 (0.17, 12)	21.4 (0.30, 9)	59.26	0.001
1990	2.8 (0.27, 10)	6.6 (0.44, 7)	4.74	0.05
1991	3.4 (0.23, 10)	10.7 (0.36, 7)	12.29	0.003

NOTES:
The values are geometric means (km), derived by back transformation from the log. distribution. Sample size and SD (× or /) are also shown.

Asami and Haga (1983) found that Brown Dipper juveniles in Japan remained with their parents until their parents started moulting in July, when the family dispersed. These juveniles are grey-brown above with numerous whitish spots; the flight feathers and coverts are tipped with white, and the feathers of the underside are whitish tipped with brown, giving the bird a mottled appearance (Ali and Ripley, 1983).

Juvenile American Dipper in Canada

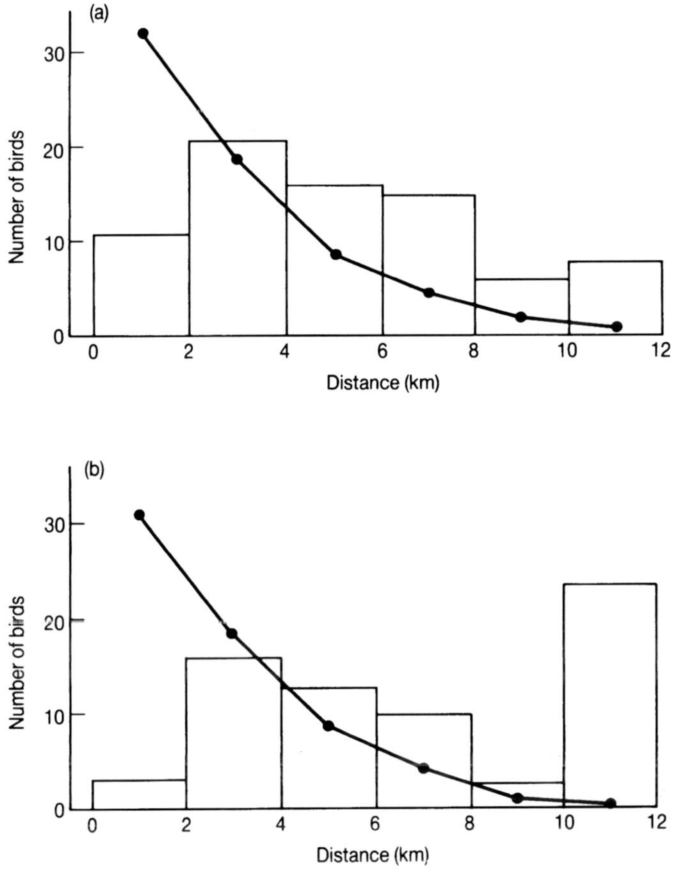

Figure 55. Actual distances moved by Welsh Dippers during natal dispersal and those expected (●) if juveniles moved far enough only to replace an annual adult mortality of 47%.

NOTES: (a) Males ($n = 77$, $x^2 = 81.49$, $P < 0.001$); (b) females ($n = 69$, $x^2 = 431.22$, $P < 0.001$). For this analysis, distances were calculated using the shortest route along the river corridor between the point of ringing and recovery (i.e. not straight-line distances travelled along the river) (Tyler, Ormerod and Lewis, 1990).

MOULT IN ADULT DIPPERS

Bird feathers are designed to be light enough to permit flight, but their resulting fragility means that they rapidly become worn and ineffective. Birds must therefore change their feathers at intervals, usually annually. In many species, moult of feathers can be expensive in terms of the materials required for growth of new feathers or in terms of loss of flight and performance, and greater vulnerability to predators. Fitting moult into the annual cycle is of major evolutionary significance, with ecological dimensions determining the moult strategies (the how, when, where and speed of moult) involved. Aquatic birds at least have the advantage of having a food supply which is available without flight. At the same time their wings often have a role in swimming or in insulation which can restrict the moult pattern. Dippers provide a beautiful example of how these conflicting restraints are satisfied.

Most adult Dippers start to moult in early to mid-June when they have finished (or almost finished) rearing young. Occasionally moult may begin as early as late May or not until August. During moult, which lasts on average about 70 days but sometimes for almost 100 days (Galbraith, Mitchell and Shaw, 1981), the birds undergo a change in behaviour. Song is rarely heard at this time, and the adults become very skulking and hard to find. They hide in crevices in rocks or in tree roots at the water's edge, being reluctant to fly.

Moult generally follows a regular pattern in the Dipper beginning, as in most passerines, with the dropping of the innermost of the ten primary flight feathers and proceeding outwards towards the tenth. Once the innermost three primaries have been replaced and growth of the adjacent two is almost complete, then the next primary is shed, although some birds shed the sixth primary before the first three are fully grown. The remaining primaries are dropped in sequence when the preceding feather is about one third grown.

Moult of the tertial feathers starts with the dropping of the fifth or sixth primary. Observation by Galbraith, Mitchell and Shaw (1981) and in Wales show that the middle tertial is shed first, followed by either the inner or outer tertial (usually by the outer), although occasionally all tertials are shed together. Richter (1954b) found that Dippers in Germany dropped the innermost tertial first, followed by the middle and then the outer feather. Tertial moult lasts about 30 days and is usually complete before secondary moult begins.

Secondary moult in Scotland started in early July (Galbraith, Mitchell and Shaw, 1981), but over 80% of 17 birds examined in Wales from 1 to 16 July in one year showed no sign of secondary moult. One late bird had only just dropped the first secondaries when caught on 18 August. The moult of the six secondaries shows much individual variation but generally the outermost secondary is shed first at the time of the dropping of the seventh primary. Then the next primary is dropped, at about the same time as the ninth primary, but further secondaries may be dropped either in order or in an irregular sequence, with the innermost one or two dropped before the third. Secondary moult continues for up to three weeks (mean 15 days) after primary moult is completed (Galbraith, Mitchell and Shaw, 1981).

Tail moult is much more rapid than wing moult, lasting about 50 days. It starts at about the same time as does tertial moult, with the dropping of the fifth primary. In many birds, all 12 tail feathers are dropped almost simultaneously; in others 2–4 of the inner pairs are shed first. Richter (1954b) found a more regular pattern with the innermost pair shed first, followed by the next pair and so on outwards.

Due to the first four or five primaries growing at the same time, the rate of moult is very rapid initially but then slows down in the latter stages of primary moult and continues slowly during secondary moult. Timing of moult varies between individuals, areas and years. Birds which only rear one brood will usually begin moulting before those that have reared two or which started breeding much later. As in Scotland, Welsh Dippers start moulting in early to mid-June with most completing their moult by the last weeks of August, but moult is spread over a wider period than found by Galbraith, Mitchell and Shaw (1981). Some Welsh birds do not start primary moult until late June and therefore do not complete their wing moult until mid- to late September, whereas Scottish birds had all finished by 1 September.

There is little difference in timing of moult between the sexes although Galbraith, Mitchell and Shaw (1981) found in one area of Scotland that males started moulting a week or two before females.

The Dipper is unique among passerines in shedding the four or five innermost primaries in quick succession. At this time it has lost some 40% in area of the wing surface. Galbraith, Mitchell and Shaw (1981) suggested that when the wings were half open, the secondaries can continue the function of the outer primaries and hence the bird can still hunt for prey underwater; they found no evidence of flightlessness in moulting Dippers. Other observers such as Balat (1960) and Snow (1967) reported that Dippers are reluctant to fly during moult, and that they seek out refuges in which to hide. Balat (1960) believed that they were actually unable to fly for a short period. Our observations support these latter authors, although we have not seen a completely flightless bird.

Dippers during moult spend much time hiding among roots and rocks at the river edge, where they forage for food. They will enter water, swimming and diving to escape capture, but are extremely difficult to flush; consequently few adults can be caught in mist nets in July. Other aquatic birds such as ducks, grebes and some rails, which can also escape from predators by entering water, also have a temporary loss of flying ability during primary moult due to simultaneous shedding of all flight feathers.

Sullivan (1965) was able to catch American Dippers when they were in moult, by running after them and capturing them by hand, or in a fish net. They tried to escape capture by hiding among tumbled logs or streamside vegetation, or by swimming away underwater. In Colorado, adult American Dippers disappeared in July from their breeding territories for about seven weeks, this absence coinciding with moult. Price and Bock (1983) found that in the summer American Dippers in their study area moved up to 25 km upstream from the breeding territories, some on to different streams, and they suggested that this movement was both to seek out good moult refuges amongst logs and rocks, and also to seek areas where food was more plentiful.

In many species of bird, moult is a stressful period when energy demands are high. Dippers, however, show an increase in mass during moult (Galbraith and Broadley, 1980; Ormerod, Tyler and Lewis, 1986) and mortality is low at this time (Galbraith and Tyler, 1982), indicating either that the moult may not be energetically stressful for these birds, or that birds had increased in weight as an insurance for this period. An investigation in Wales into whether moulting Dippers needed special prey showed that the composition of prey during moult was similar to that in winter, when the diet reflects the range of prey available (Ormerod and Tyler, 1986; Smith and Ormerod, 1986). Opportunistic foraging during moult is therefore probably an adequate dietary strategy.

Bryant and Tatner (1988) and our own studies (O'Halloran *et al.*, 1990) have also shown that daily energy expenditure by Dippers is lowest during moult, and in early winter. Moult is a time of marked inactivity, notable for the high proportion of the day spent resting and the low proportion in flight. Bryant and Tatner suggested that because flight abilities were impaired during moult, predation risk was greater when birds were feeding. By foraging rapidly, relatively more time can then be spent vigilant and under cover. The high degree of inactivity during moult may help to explain how Dippers are able to increase weight at this time.

AUTUMN MOVEMENTS AND MIGRATION

HARD WEATHER AND ALTITUDINAL MOVEMENTS

Populations of Dippers in Britain and Ireland (*gularis* and *hibernicus*) are largely sedentary, apart from local post-breeding dispersal which, as noted above, mainly affects juveniles, and females more than males. In mountainous areas where streams freeze over, or in exceptionally severe winter weather, some birds move down to open water or even to estuaries and the coast. On Scottish islands such as Rhum, and on the west Wales coast, Dippers are often seen foraging along the tideline during the winter (Tyler and Ormerod, 1986). In some inland areas such as along the Wye Valley on the border between England and Wales, Dippers may in the winter leave the small streams on which they breed and at this time birds appear on the main River Wye where few actually nest. High water levels may also cause local movements, birds being forced to fly to small streams where they are able to forage.

In central and western Europe some Dipper populations (*aquaticus*) are largely sedentary, whereas others at higher altitudes or areas subject to extreme low temperatures make regular altitudinal movements in the autumn and spring. Other populations at high latitudes where day-length is very short in the winter, undergo regular movements further south (see below). Balat (1962b) recorded Dippers as being both residents and migrants in Czechoslovakia, with the former occurring on streams fed by warm springs and hence remaining ice-free. Richter

(1953) noted that some birds behaved like migrants, leaving their breeding territories in the autumn and returning in the spring, and Kaiser (1988) too found that many Dippers in his study area in Germany migrated down to the warmer Rhine Valley, and to rivers in towns and to lakes for the winter. They occurred on some streams in very high concentrations, with territories being only *c.* 100 m. In Switzerland, Dippers also move down in the autumn to lakes and rivers outside their breeding area (Glutz von Blotzheim, 1962; Schifferli, Geroudet and Winkler, 1980), with some birds migrating across the High Alps (Jost, 1969). In Poland and Bulgaria, such altitudinal movements may also occur (Cramp, 1988). Ice-free water is crucial if birds are to remain on upland rivers for the winter. Thus in the Baltic area east to the Urals Dementiev and Gladkov (1954) record Dippers (*cinclus* and *uralensis*) overwintering at −40°C, provided that some unfrozen water remains.

There are few data on winter movements for Dipper populations in southern Europe, but Chaworth-Musters (1939) and Etchecopar and Hue (1964) found the north-west African population *minor* was largely sedentary although some birds, presumably those from the highest streams in the Atlas Mountains which freeze over, made local altitudinal migrations. There is scant information about the movements of Dippers in populations of south-east Asia (*caucasicus*, *persicus* and *rufiventris*); certainly, those further east (*cashmierensis*, *przewalski*) do make altitudinal movements, descending lower for the winter (Ali and Ripley, 1983). It seems likely that any populations breeding at high altitudes where streams may freeze over, will move downstream in the autumn.

In Colorado, Price and Bock (1983) attributed the marked 'fall' migration of American Dippers downstream, and their return in the spring, to icing over of rivers at higher altitudes. Birds that bred on rivers that remained free of ice did not move. Bakus (1959b) drew a similar conclusion in Montana, although Sullivan (1973) found that only a few Dippers were forced out by rivers icing over. Those birds that do move may fly considerable distances as instanced by one bird recorded on the Canadian Plains some 80 km from the mountains (Bent, 1948) or another seen on the north-west shore of Lake Superior in Minnesota, 1400 km from the nearest breeding habitat in the Black Hills of Dakota, by Green (1970) and Muelhausen (1970).

Migration

True migration where whole populations move away from their breeding areas is only an extension of the regular altitudinal migrations. Those Dippers breeding in Fenno-Scandia and north-west USSR (*cinclus*) and some birds from the Ural Mountains (*uralensis*) migrate distances of up to 1000 km in the autumn. Birds from Norway move mainly south-east, across the Gulf of Bothnia into Finland, and across the Gulf of Finland into Estonia and Latvia; other Norwegian birds fly south-east across the Skagerrak to Denmark (Zink, 1981). A few

Migration routes taken by Dippers of the nominate race C. c. cinlus *from Fenno-Scandia*

Scandinavian Dippers fly west, arriving on the east coast of Britain. These are the black-bellied birds occasionally recorded along the east and south coasts of England, but seen in most years in Shetland. They occur between October and April but a peak in numbers in March and April suggests that these are passage birds returning north from wintering grounds elsewhere (BOU, 1971). One such bird, ringed in Sweden in 1985, was caught two years later in early April 982 km west in Fife, Scotland (Mead and Clark, 1988).

Preuss (1959) recorded Dippers as fairly common winter visitors in Denmark, mainly in the north-east. They arrived in October, leaving again in March and April, and that they migrated at night is evident from birds being killed at lighthouses in the Kattegat area. Preuss found from museum specimens that twice as many males as females wintered in Denmark. In a wintering area in south-east Sweden, Andersson and Wester (1973) found, by contrast, that females greatly outnumbered males during the winter and on passage, the difference caused mainly by the paucity of first-year males in the area. Possibly males pass further south than females but this is contrary to the dispersal pattern that we have shown in young Dippers, and to the migration pattern in most passerines where females fly longer distances than males.

Migrant Dippers were of low weight on arrival in their winter quarters in south-east Sweden (Andersson and Wester (1972), and at their highest weights just before departure north to their breeding territory. This suggested some pre-migratory fattening as occurs in other long-distance migrants. Lehikoinen and Hakala (1988), who studied weight variation in Dippers in south-west Finland where the species is a winter visitor, found that from early March birds started to gain in weight. They showed a greater degree of pre-migratory fattening than in southern Sweden, with their April mean weight being 14–20% higher than in early March. Adult males gained weight earlier and faster than other groups and adult females earlier than first-year birds. Lehikoinen and Hakala calculated that birds put on more fat than needed merely to cross the Baltic Sea, deducing that they could fly non-stop for at least 700 km which would take them to their breeding quarters in Norway.

Sparrowhawk chasing Dipper

CHAPTER 8

Population dynamics of Dippers

Changes in Dipper populations from year to year, or from place to place, involve the interplay between birth rates, death rates, dispersal and the limits imposed by the carrying capacity of rivers. In previous chapters we have already described some of these factors in outline, but here we are concerned with the way in which they interact dynamically to maintain populations at stable levels, or cause them to vary. There are, indeed, examples of both these patterns in Wales: annual breeding surveys on some rivers show only relatively small changes in Dipper abundance from year to year. For example, we found that of 30 breeding sites checked regularly in south Wales since 1985, some 70–90% were occupied in any one year over the ten-year period. On two tributaries of the R. Usk between 50 and 70% of up to 65 sites surveyed since 1981 were occupied each year (J. M. S.

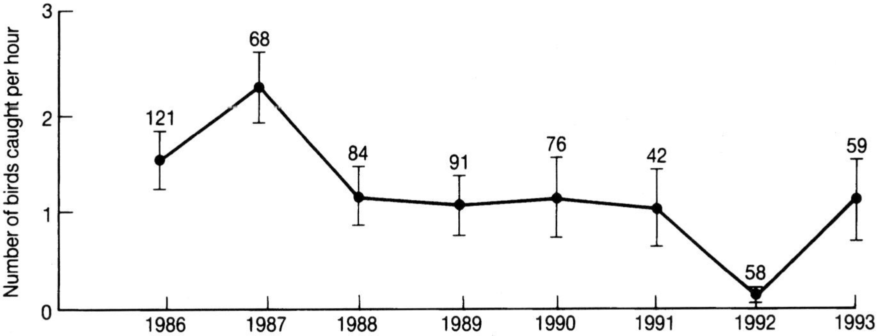

Figure 56. General stability of Dipper populations from year to year as shown by the mean catch per hour of Dippers in mist nets during annual autumn 'Dipper weekends' in mid-Wales from 1986 to 1993.

NOTE: Numbers of sites where netting took place each year is shown over the vertical bars.

Lewis, pers. comm.). Likewise autumn censuses through netting and ringing (Ormerod *et al.*, 1988b) show general similarity in Dipper abundance between years (Fig. 56). So too, the BTO's Waterways Bird Survey (WBS) indicates reasonable annual consistency in Dipper numbers at least by comparison with

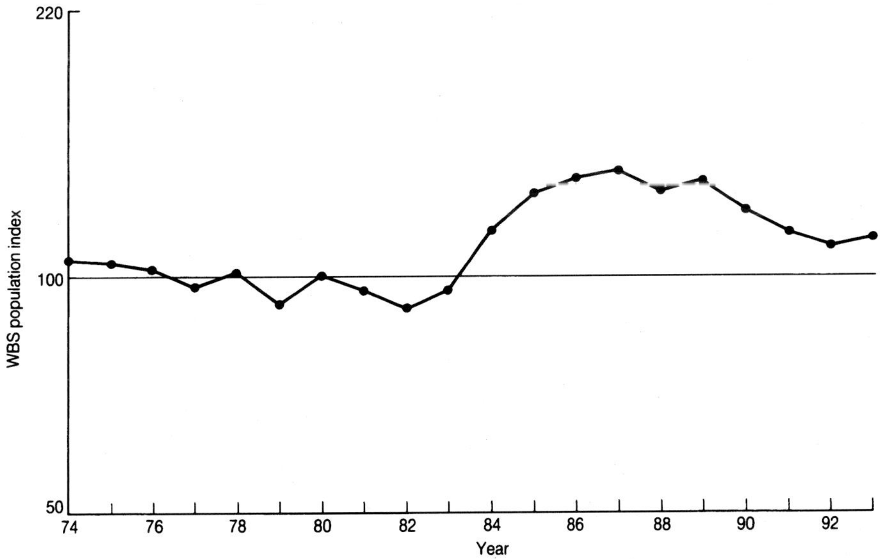

Figure 57. Annual changes in the Dipper population in Britain as shown by annual indices derived from the results of the BTO Waterways Bird Survey from 1974 to 1993 (Data supplied by the BTO).

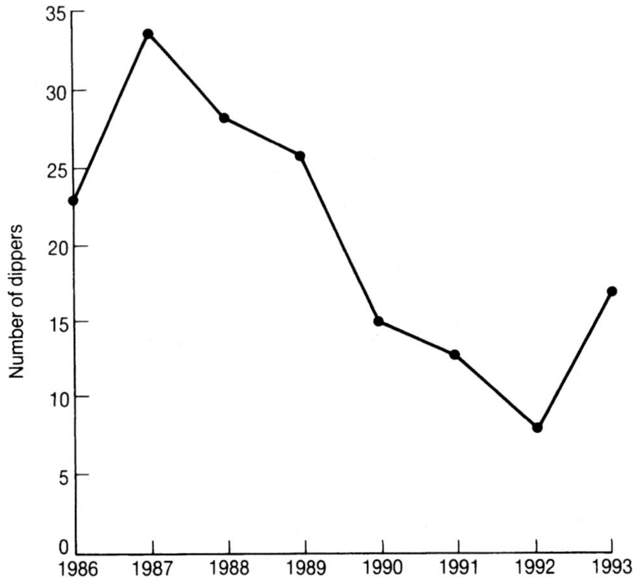

Figure 58. Decline in the Dipper population on the River Edw between 1986 and 1993 as shown by numbers caught during the annual autumn weekends of intensive trapping at roosts and in mist-nets.

other riverine species (Fig. 57). However, we know that within these aggregate patterns there are instances of population decline. For example, abundances along one apparently good quality river in mid-Wales, the R. Edw, have fallen between 1986 and 1993 (Fig. 58) even though adjacent rivers show no such trend. Abundances have also fallen on rivers such as the R. Irfon over three decades due to a reduction in water quality (see Chapter 3), whilst the acidification of rivers throughout much of north and west Wales and other areas of upland Britain has almost certainly reduced Dipper numbers over the last century or more, trends outside the time frame of any existing census programme.

POPULATIONS AND META-POPULATIONS

Conventionally, animal populations have been seen simply as the number living in a single homogeneous unit within a given area. However, recent years have seen increasing recognition of the concept of 'meta-populations', in which local semi-autonomous populations, separated by areas of unsuitable habitat, are linked through dispersal. Dippers are likely to be separated in this way because dispersing juveniles only infrequently cross watersheds between river systems (see Chapter 7). Moreover, the evidence from our Welsh studies is that adjacent rivers are linked far more frequently by such inter-basin movements than others (Fig. 59, Table 16). Rivers, or sections of rivers, which are unsuitable for Dippers, either because of natural habitat characters or pollution, might further isolate occupied systems. Dippers seldom, if ever, move from geographically distinct areas, for example the islands of Ireland or Britain, or the uplands of Wales, south-west England and the Pennines, because they disperse over such short distances. These features create the pattern through which population changes in Dippers could, under some circumstances, show geographically contrasting trends similar to those which must have occurred along adjacent acidic and circumneutral rivers.

DENSITY DEPENDENCE AND INDEPENDENCE

As with any population study, one of the keys to understanding Dipper population consistency or change is an assessment of the relative effects of density-dependent and density-independent factors. On the one hand, density-dependency involves constraints on population size as populations increase: reduced territory size, egg production, brood size or survival, and/or enhanced mortality and emigration. By contrast, density-independent factors can act equally at all levels of population density, for example during extreme weather conditions which increase mortality and reduce breeding output irrespective of how many birds are affected. In reality, a full quantification of the effects of either type of factor on any bird population would provide a long and arduous task, although important advances are currently being made through the use of *k*-factor analysis (e.g. Moller, 1989; Baillie and Peach, 1992). In such analysis, population size, age structure, breeding output and age-specific survival are all estimated annually from large samples of birds, for example through integrating several types of data collected by the BTO. Our own work on the Dipper has not yet permitted such an analysis, partly because of the difficulties of making all the correct measurements sufficiently intensively at the extensive geographical scale of our study; birth rates are well known, but rates of survival, emigration and immigration rather less so. The geographical complexities outlined above also pose a problem of understanding when a young bird has died or simply dispersed into a different river catchment. Understanding the dynamics of populations and meta-populations of animals like the Dipper becomes particularly difficult when adjacent rivers show contrasting trends in many important parameters; this is the case along acidic and circumneutral rivers. In this chapter therefore, we can only speculate about population dynamics of Dippers in Wales, and illustrate only some of the processes which might be important. Price and Bock's (1983) intensive study of American Dippers was limited by some of the same constraints.

At a larger geographical scale, however, we can make an aggregate assessment

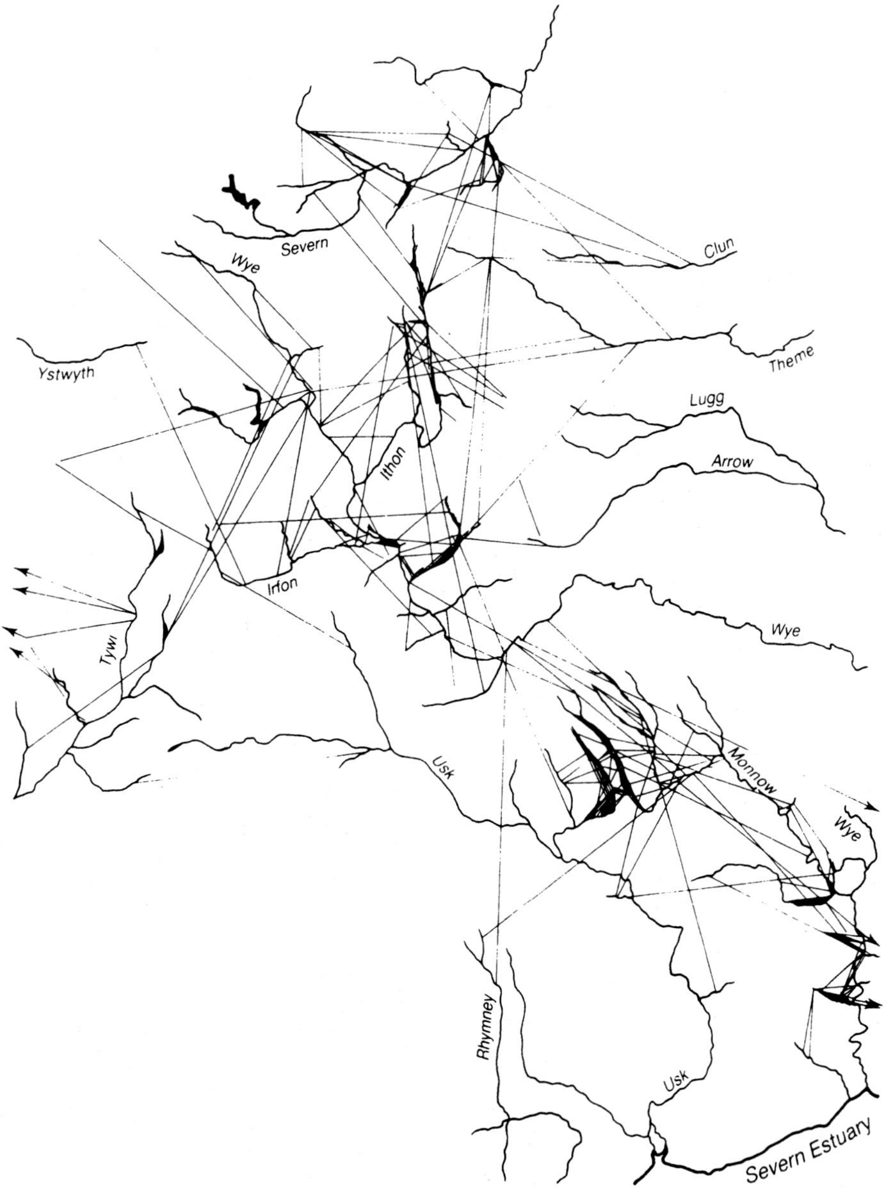

Figure 59. Meta-populations (see p. 156) as shown by the movements of Dippers, ringed as nestlings or as juveniles, on or between the main river systems in our study area in mid- and south-east Wales.

Table 16. Movements within and between river systems in mid- and south Wales made by Dippers ringed as nestlings or in their first summer/autumn (1977–93)

TO:	*Upper Severn*	*Ystwyth/ Rheidol*	*Theme/ Clun*	*Ithon*	*Irfon*	*Upper Wye*	*Mid-Wye tribs.*	*Black Mnts.*	*L. Wye tribs.*	*Usk incl. Grwyne, Fawre and Fechan*	*Forest of Dean*	*Tywi*	*Teifi*	*S Wales valleys*	*Lugg and Arrow*
FROM:															
Upper Severn	29		5	4		1									
Ystwyth/Rheidol		1													
Theme/Clun	1		30+				3								
Ithon/Clewedog/Aran	1		2	39			2								
Irfon		1			5	2	1			1					
Upper Wye		1	2	1		9							1		
Mid-Wye tribs.				2	1		66	1			2			1	
Black Mnts.							3	45	1	5	1				
L. Wye tribs.								1	42		2				
Usk incl. Grwyne Fawr and Fechan								15	2	48+					
Forest of Dean									4		20+				
Tywi					3	2				1		14	5		
Teifi												1	5		
S Wales valleys								1							
Lugg and Arrow							1								

of birth rates, adult survival and density and, by assuming emigration is roughly balanced by immigration, make a calculation of post-fledging survival. Supplementary data are also available from the BTO ringing scheme (Galbraith and Tyler, 1982). The resulting values are shown in Table 17 for both American Dippers in the Front Range of Colorado and Dippers in Britain. At these values, populations of the latter species would be increasing since post-fledging mortality of 64% is moderately below the 74% of 3.6 fledglings per adult female which would keep the population stable at 47% annual adult mortality. By contrast, the population of American Dippers would be declining since post-fledging mortality is greater than the 70% of 3.0 fledglings per adult female required to balance annual adult mortality of 45%. Of course, these estimates are subject to statistical errors, and also, in the White-throated Dipper example, they are a mix of local and national data which would not be expected to be related by the same dynamics. They illustrate, nevertheless, some of the parameters to which population dynamics in dippers are sensitive.

Evidence of effects on Dippers in Britain by purely density-independent factors such as extreme weather, is weak. Whereas other riverine bird species such as the Kingfisher and the Grey Wagtail (Fig. 60) have shown big population reductions following the harsh winters of 1961/62, 1962/63, 1978/79, 1981/82 and 1984/85 (Dobinson and Richards, 1964; Marchant *et al.*, 1990), there are no such marked effects on Dippers. Dippers in Britain maintained a stable population from 1974 (the start of the WBS) to 1983, with a slight increase since then (Fig. 57). Wide tolerance of environmental temperatures (e.g. Murrish, 1970b; see Chapter 2), as would be expected from the altitudinal and latitudinal range of the genus *Cinclus*, almost certainly protects them against cold weather. Hill streams in Britain infrequently freeze completely and in this event Dippers can move downriver or to the coast (see Chapters 5 and 7). Even in severe British weather, therefore, there is ready access to food supplies for most Dippers,

Table 17. Parameters important to population dynamics in White-throated Dippers C. cinclus *in Britain and American Dippers* C. mexicanus *in Colorado*

	White-throated Dipper	*American Dipper*
Mean clutch size[1]	4.7	4.3 (0.8)
Mean brood size[1]	4.1	2.7 (1.8)
Fledglings/adult female/year[2]	3.6	3.0
Fledgling mortality %	64.0[3]	72.1
Adult mortality	47.0[3]	45.2

NOTES:
Sources: Galbraith and Tyler (1982); Price and Bock (1993); this study.
[1] Means for all breeding attempts.
[2] Adjusted for nest failure rate and the production of second broods.
[3] Data from the BTO ringing scheme; other data from *C. cinclus* are from Wales.

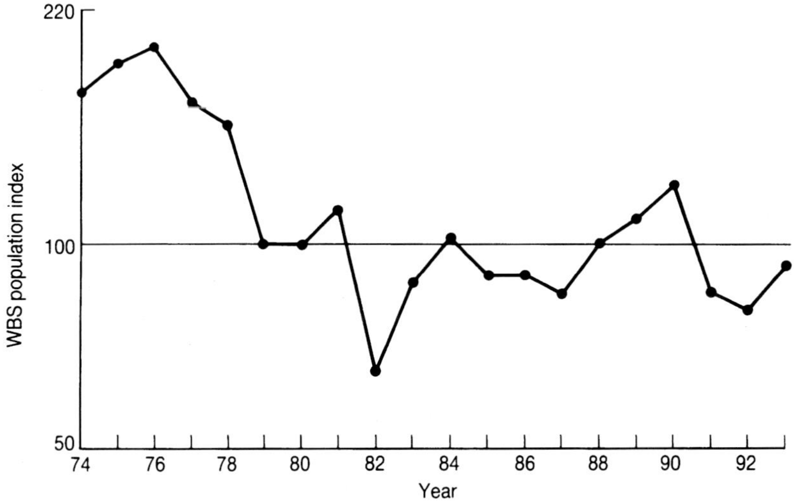

Figure 60. Annual changes in the Grey Wagtail population in Britain as shown by annual indices derived from the results of the BTO Waterways Bird Survey from 1974 to 1993 (Data from the BTO).

although in extreme winters as in 1962/63 some local populations may be adversely affected (Marchant *et al.*, 1990).

Adults show greater mortality during the period immediately following breeding than during the winter (Galbraith and Tyler, 1982), a time when body weights also reach their lowest level (Ormerod, Tyler and Lewis, 1986). There was some evidence that heavy flooding in the July and August of 1988 led to an unusual reduction in Dipper populations by the ensuing autumn. Similarly, drought in the subsequent springs and summers of 1989 and 1990 led to reductions in the contribution of first-year birds to the population in mid-Wales (Fig. 61) (possibly through enhanced vulnerability of nests to predation and/or to reduced food supplies). However, we detected no deficit in the occupancy of nest sites in the subsequent breeding seasons. That recruitment was poor in these two drought years was also evident from the small number of recaptures in subsequent years of birds ringed as nestlings in 1989 or 1990, compared with birds ringed prior to the drought years.

Mechanisms which can operate to buffer fluctuations in the breeding population in such circumstances include the surplus production of juveniles, and resulting changes in their over-winter survival because of alterations in the availability of vacant territories. So far, seasonal trends in the survival of first-year Dippers have not been widely described, although there is evidence that males which are in better 'condition' in the late autumn have improved chances of over-winter survival (Newton, 1993). Variations in the demographic make-up

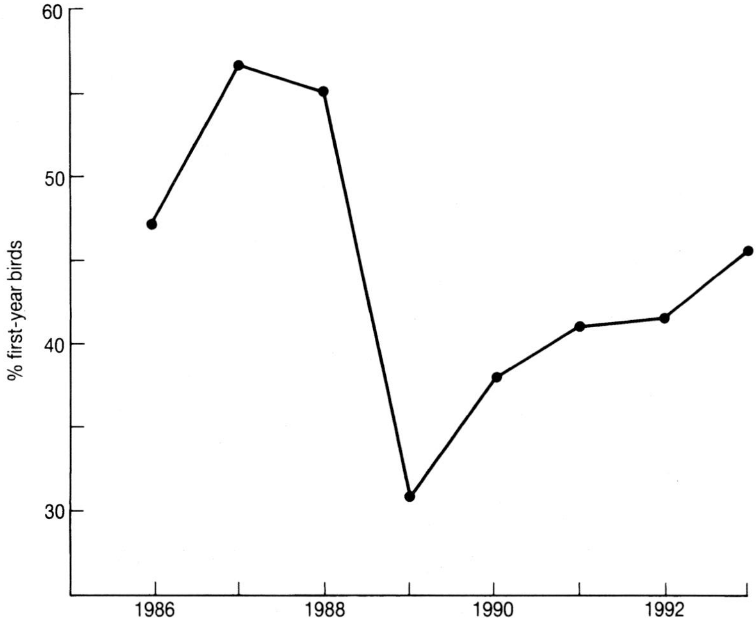

Figure 61. Percentage frequency of first-year birds among the total number of Dippers caught during annual autumn 'Dipper weekends' from 1986 to 1993.

of the wintering population (Fig. 62), coupled perhaps subsequently with facultative polygyny, might also buffer fluctuation in sex ratios (see Chapter 5); our autumn ringing data have, for example, shown an excess of males in some years and in others a preponderance of females. In these cases, it is evident that the effects of density-dependent and -independent factors, such as weather, are difficult to separate, as in other species.

Deterioration in habitat quality, such as through river pollution, are density-independent, but the overall effect is liable to be a reduction in carrying capacity along a given reach or river system as the abundance of food or other resources declines. Here, again, there will be a link with density-dependent factors which cause populations to stabilise at new, lower levels. Several dynamic processes can be involved in causing such reduction and, in the case of a change such as acidification, we are still some way from understanding which are the most important in isolation or combination (Table 18). We need to quantify not only dynamic processes within acidified areas, but also the meta-population dynamics of production and contact with adjacent unacidified areas.

Although density-dependent restrictions can impinge on the breeding performance in other bird species (e.g. Alatalo and Lundberg, 1984; Moller, 1989), we have no evidence that this is so in the Dipper. Indeed, birds at the highest

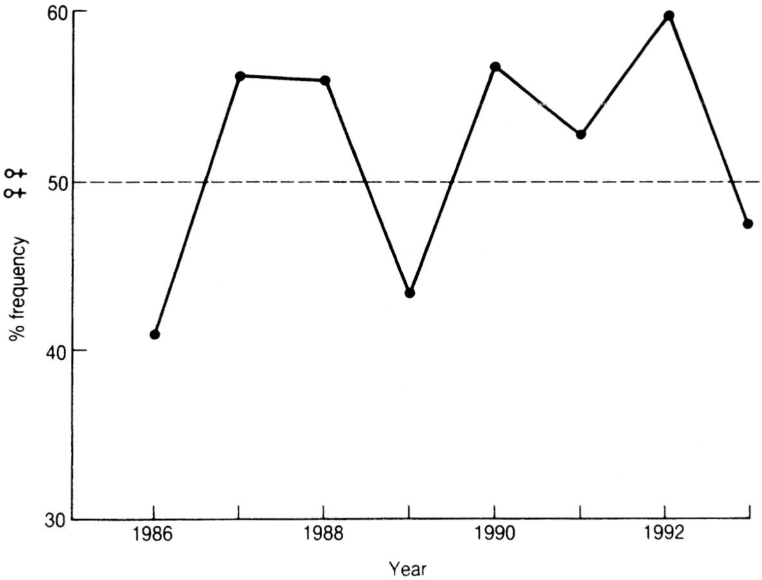

Figure 62. Percentage frequency of females in the total catch of Dippers during annual autumn weekends from 1986 to 1993.

densities in our study area breed earliest; they lay the biggest clutches and are more likely than birds elsewhere to rear second broods. These patterns are contrary to those that would be expected if breeding output were restricted at high population density. In any case, the production of young from broods of four and five nestlings (the commonest occurring) are very similar (Ormerod and Tyler, 1993c), so that any small fluctuations in brood sizes caused by changes in density would have little consequence for recruitment.

Curiously, there is some evidence of fluctuations in the breeding performance

Table 18. Population processes likely to reduce Dipper populations where habitat quality declines, such as through acidification

1. Reduced natality*
2. Reduced recruitment (i.e. reduced post-fledging survival)
3. Enhanced adult and first-year mortality
4. Reduced immigration/colonisation from adjacent areas
5. Enhanced emigration/colonisation of new sites
6. Increased territory size*

*These are known to occur.

of Dippers in south-east Wales over time, with clutch and brood sizes slightly but significantly smaller now than in the late 1970s or early 1980s (see Chapter 6; and Table 19). We have no direct measure of how these changes reflect population density, nor how they have affected it, but there is no evidence that populations have either decreased or increased as these changes have occurred. Crick, Dudley and Glue (1993) found too from BTO nest record card analysis that there had been an increased failure rate during the nestling period of Dippers in recent years, although they found that because this was offset by lower losses during incubation during this same period no overall increase was evident in nest failure (egg laying to fledging).

Most likely, the strongest density-dependent effects on the population size of Dippers is liable to arise through territoriality. Territory size is reduced strongly where populations increase (see Chapter 3), with gaps in the pattern of distribution seldom sufficiently large to accommodate new settlers. Moreover, exploitation of food per unit area by Dippers is large (1–2 g dry mass per m^2 per year; Ormerod and Tyler, 1991a), so that the defence of sufficiently large territories must be important in guaranteeing year-round food supplies. As further evidence that food involved in setting territorial limits, and hence carrying capacity, territories increase in size where food is scarcest (see Chapter 3). While we have insufficient data to calculate variation in territory size from year to year, observations indicate that it cannot be large. Density and nest occupancy are generally predictable from year to year, so that here is at least one likely source of strong regulation which keeps the population within limits. The fate of young birds or displaced adults which are excluded from occupied territories is unknown, but must involve either further dispersive movement or death since there is little evidence of a surplus non-breeding Dipper population. Again, however, we have no measures of how these patterns of movement or mortality relate to Dipper population dynamics.

Defence of territories containing roost sites appears less critical, since Dippers will venture outside their own territory to roost in adjacent or nearby territories. Nest sites will usually be available, at least on upland rivers.

Table 19. Long-term trends in breeding performance of Dippers in south-east Wales

	Clutch size	*Brood size*
1976–81	4.85 (0.61, 133)	4.29 (0.93, 151)
1982–87	4.69 (0.61, 163)	3.98 (1.03, 259)
1988–93	4.66 (0.57, 124)	3.98 (1.02, 193)

The values are means with standard deviation and sample size in parentheses. The reductions in clutch and brood size over time are statistically significant.

AMERICAN DIPPERS IN THE FRONT RANGE OF COLORADO

The most intensive study of dipper population dynamics conducted to date involved American Dippers in the Front Range Mountains of Colorado. Price and Bock (1983) ringed 558 individuals on and around 30 km of rivers in a two-year period, subsequently measuring densities, reproductive performance, dispersal and survival in relation to environmental factors. They specifically selected American Dippers for study because of the species' 'simple' ecology and

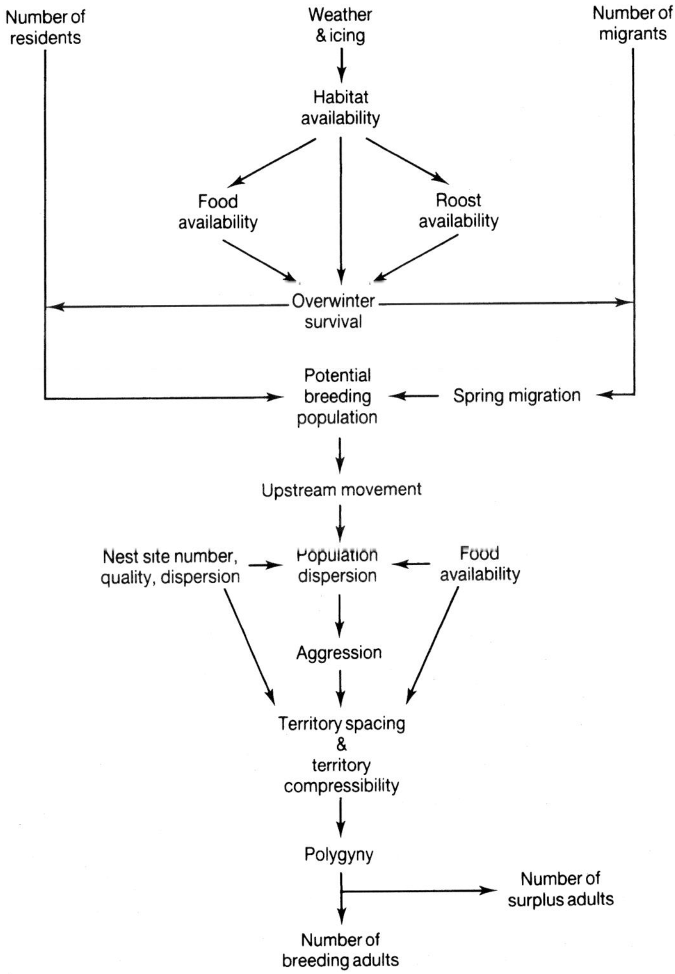

Figure 63. Population dynamics in the American Dipper (Price and Bock, 1983). The figure shows factors suspected of having effects on population size.

ease of counting and observation, especially well suited for population dynamics. Even so, their study led to difficulties in adequately separating survival from emigration, and allowed only correlative evidence of factors liable to influence or regulate population. Seventeen important factors were identified, among which density independence and dependence interacted in complex ways (Fig. 63). Moreover, different processes were important under different circumstances: if breeding success had been poor due to shortage of food or nest sites, then food or ice-free rivers might not be in short supply in the following winter. By contrast, if over-winter survival were low due to low food abundance or excessive icing over of rivers, then territoriality might not limit breeding density the following spring since there would be sufficient resources for all birds.

For the time being, therefore, we have to be content that 'average' values for adult survival, post-fledging survival, or any other factor which regulates dipper populations, hide a wide range of variations and interacting processes of which we still know surprisingly little.

Dipper and Grey Wagtail

CHAPTER 9

Relationships between dippers and other species: competitors, predators and parasites

Wherever any of the five species of dipper occur, they come into contact with other birds, mammals or other vertebrates using the same habitat. Some of these animals may be potential competitors for space, food or even nest sites; others may be coexisting with the dippers, utilising different parts of the same habitat or preying on different items; yet others may be predators. Invertebrates as well as providing much of the dippers' food, may also parasitise the birds, internally as well as externally.

INTERSPECIFIC COMPETITION AND COEXISTENCE

DIPPERS AND WAGTAILS

One species of bird that is frequently found on those streams and rivers favoured by Dippers in Britain and western Europe is the Grey Wagtail. Although Grey Wagtails have a wider distribution, pairs of both species often breed within the same territory and select similar nest sites. Nests of Dippers and Grey Wagtails may even be adjacent to each other under the same bridge. Generally there is little aggression between the two species.

As both species prey on aquatic insects, competition could be expected. However, larval and adult flies (Diptera) are important prey for Grey Wagtails, often derived originally from terrestrial vegetation and trees (Ormerod and Tyler, 1987a; 1991b). Grey Wagtails do take some caddis and mayflies but it is generally the adults rather than larvae or nymphs that they catch by flying out over the water. They will also peck items from rocks, from the water surface or along the edge of the river, where they wade, but they cannot dive. They may forage a short distance away from the river corridor on moorland and pasture or on paths and tracks (Ormerod and Tyler, 1987a; 1991b).

Grey Wagtails are therefore more opportunistic than Dippers, have a wider range of prey and different foraging techniques. These differences might allow both species to coexist on the same stretches during the breeding season. In the non-breeding season, Grey Wagtails are less dependent on fast-flowing streams, often feeding well away from water at farms, on lowland rivers, in towns at sewage works, on woodland tracks and other such sites.

Sometimes, where nest sites are limited, there may be direct competition for sites between Dippers and Grey Wagtails. In eastern Germany, Kaiser (1988) found that Dippers and Grey Wagtails competed for nest sites but that Dippers were usually dominant. This appears to be the case in Wales where we have recorded Dippers taking over part-built nests of Grey Wagtails. At lower altitudes, Dippers start breeding 2–4 weeks before Grey Wagtails and hence most are already incubating or have young by the time Grey Wagtails start nest-building (Fig. 64). At higher altitudes though, both may start building at the same time, and competition for sites could occur. Grey Wagtails sometimes build a nest within a Dipper's old nest. Where the two species do nest close together, one may feed the brood of the other species, particularly if the former has lost its own young (see Tyler, 1972). Yoerg and O'Halloran (1991) watched a male Grey Wagtail in mid-Wales repeatedly taking food to nestling Dippers, even though the Wagtails had small young of their own.

Despite this coexistence between the two species, there are occasional and brief aggressive interactions, with either of the species being the aggressor. Eggbrecht (1937) found that male Dippers often chased Grey Wagtails from the vicinity of the nest. We have seen such encounters and also noted juvenile

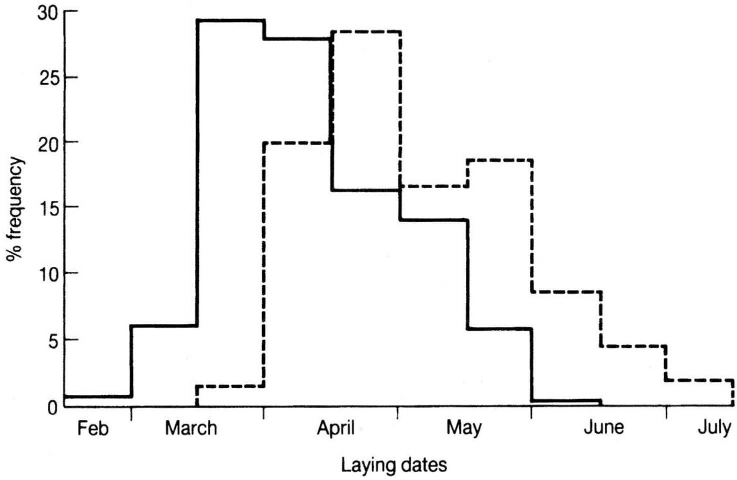

Figure 64. Comparison of the breeding season of Dippers (solid line) and Grey Wagtails (dotted line) in Wales as shown by the distribution of first egg dates for 1263 Dipper clutches and 721 Grey Wagtail clutches for the years 1978–93.

Dippers sometimes chasing Grey Wagtails. In the Himalayas, David Pritchard (pers. comm.) observed a Grey Wagtail aggressively approach a Brown Dipper on a mountain river; the latter responded by diving, then resurfacing and flying away from the Wagtail. The significance of these aggressive interactions is not known.

Although the breeding distributions of the two species overlap considerably in upland areas of Britain, Grey Wagtails breed too in counties of southern and south-east England where Dippers do not occur as breeding birds. Grey Wagtails will even nest by slow-moving rivers provided these have weirs, mills and associated mill races or areas of faster-moving clear shallow water.

Breitenmoser, Wursten and Marti (1987) found that both species occurred together in Saaneland in Switzerland but that Grey Wagtails bred on smaller and on higher streams than did Dippers. Grey Wagtails were found up to the timberline or even above it to 1780 m, whereas Dippers only bred up to 1590 m. Surprisingly, in view of the differences in their diet and choice of feeding areas where they coexist, Grey Wagtails occurred at highest densities on the higher watercourses where Dippers were absent.

On some Dipper rivers, particularly in north-west Britain, the Pied Wagtail (and in Europe the White Wagtail) may also be found. As with Grey Wagtails, differences in diet and foraging behaviour may permit coexistence although some competition may occur for nest sites. Pied Wagtails are notoriously aggressive to many other birds, so it is not surprising that aggressive encounters have often been recorded between these Wagtails and Dippers. In

Northumberland on the R. Coquet, D. J. Perray observed three Pied Wagtails repeatedly attacking a Dipper which had been feeding in shallow water (Perry, 1986). As the attacks intensified with the Wagtails diving at its head, the Dipper submerged into deeper water. When it surfaced downriver, the Wagtails renewed their attack forcing it to move further away.

Dippers and Wrens, Kingfishers, Sandpipers and Other Birds

In Britain and Europe, Dippers regularly come into contact with Wrens, which when living by rivers frequently choose similar nest sites to Dippers, and nest in close proximity. The much larger Dipper is clearly dominant where preferred nest sites are limited, as in crevices under bridges. Kaiser (1988) recorded competition for nest sites between Dippers and Wrens and also between Dippers and Black Redstarts, but noted that Dippers were usually dominant. Aggression towards Wrens by male Dippers was also noted by Eggbrecht (1937), and by Rankin and Rankin (1940) who observed a Wren building its nest close to that of a pair of Dippers, and a Dipper then chasing it every time it tried to approach its own nest. The Wren, however, was undeterred and continued to build its nest. Indeed the two species frequently rear broods only metres apart. Le Hardy de Beaulieu (1973) even found a Wren that had built its nest on top of an occupied Dipper nest.

Wrens often use the old nests of Dippers during the winter for roosting. After the winter of 1981/82 we found six dead Wrens in one old Dipper nest below a bridge.

Wren inspecting Dipper nest

Occasionally Dippers nest close to Kingfishers, although the latter generally inhabit the wider slower rivers where Dipper abundances are lower. There is no real competition between these birds as they select very different prey, but sometimes they will dispute space and behave aggressively towards one another. Eggbrecht (1937) recorded Dippers chasing Kingfishers (and Song Thrushes) that approached too near to their nests. Ken Perry (1986) records seeing a Dipper in a sycamore tree over a river in southern Ireland when a pair of Kingfishers alighted and copulated on a branch higher up in the tree. The Dipper then flew to a branch close to the Kingfishers and, standing in the upright threat posture, dipped and called its 'zit' alarm several times in an excited way until the Kingfishers flew off. It is not always the Dipper that is aggressive in encounters between the two species. Grist (1934) records a Kingfisher repeatedly 'attacking' a Dipper that was feeding nearby. Each time the Dipper brought a beakful of food to the river edge, the Kingfisher flew down from its perch, landing by the Dipper. A scuffle ensued, resulting in the Kingfisher eating the Dipper's food. After the Dipper had been robbed of its food two or three times, it moved away from the Kingfisher. Unfortunately no mention was made of the type of prey obtained by the Dipper.

On those rivers where Common Sandpipers and Dippers occur together, interspecific territorial defence may occasionally be seen, with Dippers chasing sandpipers from their nest site or feeding area. Generally higher densities of Common Sandpipers occur lower down rivers where Dipper densities are low, but there are rivers in mid- and south Wales where both species occur in high densities. Here both Common Sandpipers and Dippers feed along shoals at the river edge, but Yalden (1986) has shown that adult sandpipers and their small chicks feed for much of the time away from the river in short grass and in wet flushes. Chicks older than five days up to fledging at about three weeks do feed predominantly by the river, but their diet consists largely of adult flies (Dipterans), and terrestrial invertebrates. Stoneflies, mayflies and caddis, the Dippers' main prey, comprise less than 20% of the adult sandpipers' diet and even less of the chicks' diet (Yalden, 1986). Common Sandpipers occur on rivers in Britain and central/northern Europe from April to September, so present no competition with Dippers during the winter, although in southern Europe and North Africa, Dippers and Common Sandpipers can be found together at this time.

Interactions Between Dippers and Other Species of Bird

Acts of aggression between Dippers and other species are uncommon. Ken Perry (1986) noted that in 25 years of observing Dippers, he has rarely seen confrontations between them and other species. However, he quotes D. R. Collins's observation of a Little Grebe behaving aggressively towards a Dipper. The Little Grebe was diving and feeding in a loch and a few metres up the inflow

Common Sandpiper

river, while a Dipper was searching for food in shallow water, occasionally submerging in deeper water. The grebe swam to within two metres of it with its head thrust forward. This aggressive action was seen several times with the Dipper retreating each time until it flew off upriver. In north Wales, Moorhens were observed trying to dunk a Dipper.

John O'Halloran (pers. comm.) observed a Spotted Flycatcher collecting moss for its nest from an occupied Dipper nest in mid-Wales in 1988. The flycatcher made such a large hole in the dome of the Dipper's nest that rain entered and the nestlings died.

Interactions Between (White-throated) Dippers, Brown Dippers and Other River Birds

In the eastern part of the Dipper's range in eastern Russia and in the Himalaya, it comes into close contact with a second species of dipper, the Brown Dipper (*C. pallasii tenuirostris*). Competition between these two closely related species would seem inevitable, given the marked impact on benthic fauna by dippers. Both species have been recorded breeding in close proximity in similar habitats, e.g. in Russia by Kovshar (1979) and in Burma by Stanford and Mayr (1941), but the larger Brown Dipper is according to some observers, dominant, driving the smaller dippers from their territories. Kovshar records Brown Dippers ousting two pairs of Dippers (the white-bellied race) from their nests in subalpine Tien Shan in the USSR, and Stepanyan (1962) found no Dipper within a Brown Dipper territory. Budris (1981a) though, saw no aggression between the two and found nests of the two species in different sections of the river system. He concluded that there was no competition. However, he found only one Dipper nest, and the segregation of breeding territories of the two species would imply some exclusion was occurring. Budris (1981b) later stated that the larger Dipper forces the smaller Brown Dipper from its nesting area. (In fact it is the Brown Dipper that is the larger.) He suggests the Brown Dipper is ecologically more 'plastic', nesting in more varied types of habitat, whilst the Eurasian species is confined mainly to river sites and streams with high gradients. On this he is in agreement with others.

Most authors do generally agree that where the ranges of the two species overlap, the White-throated Dippers breed at higher altitudes and on smaller streams with steeper gradients than their larger relative (e.g. Kovshar, 1979; Ali and Ripley, 1983; Tony Gaston, pers. comm.; Andrew Harrop, pers. comm.). In Nepal where the former species is scarce, the Brown Dipper then occurs on smaller rushing glacier torrents up to 5000 m, at higher altitudes than elsewhere. Mallon (1987) found during the winter months that Brown Dippers were more abundant than Dippers (3:1) in Ladakh, the former occurring on hill streams as well as on major rivers but the latter mainly on hill streams.

It is surprising in view of the consensus of opinion outlined above, that Shul'pin (1965), quoted in Budris (1981b), suggested that as the Brown Dipper starts breeding earlier than the white-bellied Dipper, it can occupy optimal areas in the *upper* parts of river valleys, leaving the white-bellied Dipper with less favourable areas further *downstream*.

Budris looked at the two species in the Mountains of Badakhshan, and found that Brown Dippers nested on fast-flowing mountain rivers with rock walls and boulders whereas the white-bellied Dippers bred in an overhanging peat bank on a slower-moving braided section of river. His sample size of nests, however, was very small, as was that of Stepanyan (1962) in Tien Shan, who found differences in nest site location between the two species. More detailed observations are needed on nest sites, nest construction, diet and behaviour of the two species, before comparisons can be made between them.

The two species of Dipper in the Himalayas occur alongside many other river birds, notably Blue Whistling Thrushes, White-capped and Plumbeous Water-Redstarts (formerly known as River Chats and Plumbeous Redstarts) and at least three species of the beautiful forktails (Little, Spotted and Slaty-backed, and in the winter, Black-backed too), as well as Grey Wagtails. Little is known of the ecological segregation of these various river birds in the breeding season. Stepanyan (1962) recorded a pair of Brown Dippers nesting within 2.5 m of a pair of Blue Whistling Thrushes for eight years. Budris (1981a) claimed that there was no competition for food between Brown Dippers and two other species frequenting streams in the mountains of Badakhshan. The White-capped Water-Redstart foraged on the river banks and adjoining slopes, the Brown Dipper on the bed of the stream and the Little Forktail took aquatic prey on stones washed by water in the middle of the stream.

Pilot surveys that we carried out in November and December of 1989 and 1991 and in March 1992 into the ecological segregation of river birds in the middle hills and Himalaya of Nepal generally confirmed Budris' findings. We looked at nine species of river bird including Brown Dippers, and showed that their size, the use of the river corridor and foraging techniques differed greatly between the species, so allowing them to coexist during the winter months (Tyler and Ormerod, 1994). The large Brown Dippers took prey from submerged rocks and the river bed; the small and short-tailed Little Forktails took prey from spray-drenched rocks by waterfalls or from the wetted areas of boulders in the river; Plumbeous Water-Redstarts caught insects mainly by aerial fly-catching; the larger White-capped Water-Redstarts caught much prey in the air too but also spent much time picking prey from shoals and from the riparian zone, as did Grey Wagtails; the long-tailed Spotted and Slaty-backed Forktails turned over stones, leaves and other debris in the marginal and riparian zones. Both Slaty-backed and Black-backed Forktails also foraged among alders, on moist tracks and in riparian vegetation. Brown Dippers were rarely aggressive towards the other species, whereas Plumbeous Water-Redstarts, despite their small size, were the most aggressive birds.

The American Dipper and Other River Birds

In North America, there are few species of birds on 'dipper streams' to be potential competitors (Price and Bock, 1983). There are no wagtails although flycatchers, such as the Black Phoebe, live alongside 'dipper' rivers from California south to Argentina, preying on aerial insects by and over the water. The Louisiana and Northern Waterthrush occur by small streams and swamps in forests, and prey on aquatic invertebrates, notably caddis (Trichoptera) larvae in the case of the former species, and dipteran larvae in the case of the latter (Craig, 1987). In view of differences in foraging behaviour between the American Dipper and the Black Phoebe and two waterthrushes, and the type of wooded wetland chosen by the waterthrushes, competition is unlikely.

On many American rivers, Spotted Sandpipers take the place of the European Common Sandpiper. Where American Dippers occur on these slower pebbly reaches, they sometimes pursue Spotted Sandpipers within their territory (Bakus, 1959b), although both species do breed successfully on the same stretch.

Harlequin Ducks occur on rivers in the Canadian Rockies and dive for aquatic invertebrates, such as caddis larvae and stonefly nymphs, on the river bed. Generally though, Harlequins use larger and deeper rivers than those preferred by the American Dipper. The Belted Kingfisher also feeds mostly on larger rivers than the streams favoured by American Dippers, and in a wider range of habitats, including large deep lakes. It preys almost exclusively on fish (Bent, 1940).

South American Dippers and Other River Birds

Several species of river birds occur alongside White-capped and Rufous-throated Dippers. The best-known of these is the Torrent Duck, which occurs throughout the Andes and dives for stonefly nymphs and caddis larvae in the torrential rivers; possibly the reason for the South American dippers rarely if ever diving may be because this diving niche is occupied by the much larger Torrent Ducks. Less well known is the Torrent Tyrannulet, about half the size of a dipper but similarly coloured. It nests on rock faces or even under bridges as do dippers, but catches mainly aerial prey by hovering over rushing mountain rivers. Other tyrant flycatchers and wagtail tyrants also feed on insects by and over South American mountain rivers, whilst various furnariids such as Bar-winged and White-winged Cinclodes commonly occur alongside Rufous-throated Dippers in Argentina. Salvador, Narosky and Fraga (1986) suggested that one of the reasons why the Rufous-throated Dipper is so restricted in range was because of competition with these and other *Cinclodes*. Our observations, however, indicate that although these *Cinclodes* sp. wade in water and prey on aquatic invertebrates, they forage mainly on shoals and in the riparian zone, picking amongst stones and vegetation and turning over pebbles and debris in the manner of Spotted Forktails in the Himalaya (Tyler, 1994). They do not feed on the lips of cascades and waterfalls, a favoured location for the Rufous-throated Dippers. We believe therefore that competition between these river birds is largely avoided and that it does not explain the very restricted range of Rufous-throated Dippers. Ringed Kingfishers and Fasciated Tiger-Herons also hunt along forest streams in north-west Argentina, but these birds take fish, amphibians and large crustaceans rather than nymphs and larvae of stoneflies, mayflies and caddis flies.

Competition Between Dippers and Fish?

Throughout the range of the five species of dippers, it is clear that we need to understand better the relationships between rivers and their birds, and other forms of wildlife. For example, we do not know even in Europe what the impact of fish predation is on the invertebrate community in the river, and hence whether fish and dippers compete for food. Some fish and dippers do feed on the same items, but fish, notably trout, take lots of drifting invertebrates, and input of these from terrestrial sources can be huge: 40 g/m^2 over the summer alone (Ormerod and Tyler, 1991a).

Price and Bock (1983) suggested that trout were likely to be competitors of American Dippers because of overlap of food (Carlander, 1969). They provided no data on niche overlap between Rainbow Trout, Brown Trout and Brook Trout and American Dippers in their study area in Colorado, but drew attention to potential differences between the niches of trout and dippers, such as preferred water depth, substrate, time of feeding and proportion of prey taken as drift (for references: see Price and Bock, 1983).

The effect of predation by fish on the aquatic benthic invertebrates remains controversial. Some experimental studies suggest that fish predation may control the abundance and diversity of species of aquatic invertebrates (e.g. Gilinsky, 1984; Morin, 1984); others dispute this (e.g. Thorp and Bergey, 1981).

Populations of Brown Trout and Dippers occur together on many upland rivers in Britain and Europe. Brown Trout feed on caddis larvae and mayfly nymphs, the most important prey for Dippers, as well as on terrestrial invertebrates, especially where trees overhang rivers. We do not yet know whether high stocking rates of the native species and introduced Rainbow Trout have any detrimental effect on Dippers, whether perhaps Dippers occur at lower densities on streams where trout numbers are artificially high, or the reverse where trout are few. Unfortunately, those streams in Wales where trout are scarce are the acidic streams where the most important prey for Dippers is also scarce.

Competition for caddis larvae between Blue Duck in New Zealand and introduced trout has been suggested as a reason for the decline in duck numbers (Kear and Burton, 1971), although Noda and Stanton (pers. comm.) noted that dietary overlap was small, with trout eating much larger larvae than those taken by Blue Ducks.

Competition Between Dippers and Aquatic Mammals?

Little is known either of the extent of competition between dippers and small carnivorous mammals such as the Water Shrews in Britain and Europe and a closely related species in the Mediterranean area and in Asia Minor. In Spain studies have been made by Jesus Santamarina (pers. comm.) of the relationship between Dippers and the Pyrenean Desman, a rare mammal confined to the Pyrenees. The Pyrenean Desman feeds largely on caddis larvae, mayfly nymphs and other aquatic insects and is therefore potentially in direct competition for food with Dippers, but Santamarina has shown that the Desmans take fewer hydropsychid caddis larvae, molluscs and dragonflies (Odonata) than do Dippers. Moreover, the Desmans take many invertebrates from rivers with slow-moving water, with a sandy substrate and an accumulation of dead leaves and branches, a habitat little used by Dippers.

Beavers in North America, and the rarer European Beaver, do not compete with dippers for food but they have a major impact on the dippers' habitat by felling trees and creating dams. This can considerably alter the ecology of a watercourse. A good dipper stream can be converted into a beaver pond and swamp, and traditional nest sites and feeding areas can therefore be destroyed by the beavers' activities.

PREDATORS OF DIPPERS

Dipper eggs, nestlings and full-grown birds are subject to predation by a wide variety of mammals, birds and other vertebrates. Predation may be a major cause of nest losses in some areas or in some years. Efteland and Kyllingstad (1984) found predation to be the most important factor in reducing nest success in a Dipper population in Rogaland in south-west Norway. In Colorado though, Price and Bock (1983) wrote: 'because of (American) Dippers' alertness, their open habitat, and the inaccessibility of most nests, we do not feel that predation is a major cause of mortality for adults or nestlings'. Nonetheless, they accepted that newly fledged juveniles were vulnerable, and that predation was a major cause of nest failure.

MAMMALIAN PREDATORS

Rats and other rodents

In Ireland, Perry (1986) found mammals, mainly rats, could cause heavy losses of nestlings (up to 75% of nest failures due to non-human predators).

Geoffrey Shaw (1978) in an analysis of cards submitted to the BTO Nest Record Scheme up to 1972, also found that rats accounted for a high proportion (60%) of nest failures due to predators. In our Welsh study area, rats have been suspected as predators of both eggs and nestlings at a large number of sites, especially those which by nature of their position (e.g. in drainpipes high on a bridge wall or on high ledges below bridges) are largely inaccessible to any other predator. Rats can climb well and often live along rivers, especially near houses or farms.

On one occasion when we were checking nest sites on tributaries of the River Severn, we felt for eggs or young below what we initially thought was the incubating or brooding female. Some females do sit very tight, allowing inspection of the nest contents without leaving the nest. However, the 'bird' appeared to be unusually warm and to have fur and a raised backbone so hands were hurriedly withdrawn and a torch fetched to examine the nest contents. A rat was revealed, curled up in the nest, either replete with eggs or nestlings, or perhaps moribund due to having eaten rat poison which was known to be put down in a nearby factory adjoining the stream. Only days previously when checking another nest on a tributary of the Severn we had found a large grass nest sited only half a metre below a Dipper nest containing young in the roots of an upturned tree. A rat bolted from this as soon as we touched the nest. The young Dippers in the adjacent nest, perhaps surprisingly, fledged safely. Older Dipper nestlings may be extremely noisy when calling for food and frequent visits by the adults to the nests may also draw the attention of predators such as rats. Sackl and Dick (1988) found that survival of older Dipper nestlings on the Kamp river system in lower Austria was much lower than during incubation or early stages of the nestling period; they attributed this to the older broods being more vulnerable to predators because they were so vocal.

Incubating adult Dippers may also fall prey to rats; a female known to be six years old, had hatched nestlings in a drainpipe nest in a tunnel beneath a road, but two days later she was found part eaten behind the nest, with no trace of her nestlings. No other mammal, except possibly a feral mink, could have been responsible. On Rattlesnake Creek, Missoula County in Montana, Bakus (1959a) found an adult Bushy-tailed Woodrat in an American Dipper's old nest, and suggested this mammal may be a potential threat to nesting birds.

In Poland, Czapulak (1988) recorded a Dipper brood taken by a dormouse.

Mink and other mustelids

Feral American Mink, escapes or deliberate releases from fur farms throughout Britain, undoubtedly prey on Dipper eggs, nestlings and juveniles. Mink were thought by Perry (1986) to be one of the major causes of mortality in young Dippers in Ireland, especially in the period prior to and also after fledging. Perry suggested that a decline in the Dipper population in Dumfriess-shire in Scotland since the 1970s was due to mink predation and that, if mink further expand their

Mink below Dipper nest

range, then they could become a major threat to Dipper populations; however, he presented no evidence to support this.

We have no evidence to show any adverse effect of Mink on Dippers in our Welsh study area. Mink are present on most of the rivers and yet there has been no decline in Dippers attributed to them since the start of the study in 1979 (Tyler, 1991). We know that Mink rob some nests and take fledged young but there is naturally a high mortality among juveniles with 65% dying before their first winter (Galbraith and Tyler, 1982). Populations of certain other waterbirds, notably Moorhens, may however be locally, or temporarily, depressed by Mink (Tyler, 1991).

The mink native to Europe is a natural predator of Dippers, whilst the American Mink is also a predator of the American Dipper (Steiger, 1940). Other mustelids also prey on dippers. Thus Shaw (1978) noted that stoats were occasionally recorded as nest predators. People have observed juveniles being eaten by stoats or weasels. Stephens (1957) recorded Otters taking waterfowl, especially Moorhens, and domestic ducks and poultry; nests of gulls may be raided and even adult gulls taken. Otters are widespread now on most Welsh rivers. Their holts may be very close to Dipper nest sites and it is therefore quite likely that they sometimes take nest contents or young birds. Polecats, common in Wales and the borders and elsewhere in upland Britain, no doubt prey on Dippers too. In North America, martens, skunk and weasels sometimes catch American Dippers (Steiger, 1940).

Cats

Domestic and wild cats may be predators of inexperienced young Dippers or unwary adults at some sites. Up to 1985, some 190 ringed Dippers had been reported to the BTO as found dead. Of 77 birds whose fate was known, 33 (almost 43%) had been killed by cats. Others had been brought in alive by cats but were released unharmed. In Wales since 1985 one cat has caught at least three ringed Dippers. These figures have to be interpreted with caution; a much higher recovery rate would be expected of birds killed by cats rather than by wild predators, because of the domestic cat's habit of bringing its prey into its owner's house.

AVIAN PREDATORS

Crows

Birds, notably corvids (members of the crow family) and both diurnal and nocturnal raptors, account for some losses of nestlings and full-grown Dippers. Shaw (1978) found that one nest failure out of only 12 where the predator was identified, was due to a Carrion Crow and another to a Jackdaw. Corvids have been implicated as predators of nests in many other dipper studies. In 1989 Dominic Hudson and John O'Halloran (pers. comm.) observed Jays in two instances taking Dipper nestlings in Wales. Magpies and crows have also been seen attacking nests. Predation by corvids on adults is less well recorded but Svein Efteland (1983a) observed a Magpie and Carrion Crow chasing a Dipper along a stream in south-west Norway, the Dipper only escaping by quick and sudden flight manoeuvres.

Perry (1976) found no records of nest losses in Ireland due to avian predators, crows or otherwise.

Diurnal raptors

Sparrowhawks may be the commonest bird predator of full-grown Dippers. Many fishermen and bird watchers have reported instances of this common hawk chasing Dippers and Kingfishers; we have frequently seen Sparrowhawks chasing Dippers (and Grey Wagtails) in Wales and Perry (1986) records a Sparrowhawk intercepting a Dipper as it flew to its nest with food. Tree-lined streams provide an ideal hunting situation for this small predator, and Dippers flying low over the water on a regular route offer an easy target, although the Dipper often escapes by dropping like a stone into the water. Creutz (1966) noted that when a Dipper was pursued by a Sparrowhawk, it flew low over the water in a zigzag manner, and either dropped into the water or hid under vegetation when the predator came too close.

Kestrels, Merlins and Peregrines occasionally take Dippers as some eye-witness accounts show (Joan Shotton, pers. comm.). Colin Bibby (pers. comm.) analysed bird remains at 66 Merlin sites in Wales and of over 1600 individual bird prey items, none were Dippers. It is likely though that the close proximity of breeding Merlins to Dippers in some upland valleys results in Dippers being taken as prey. There is a ringing recovery in the BTO files of a Dipper ring from a pellet in a Peregrine nest. Tony Balbi in N. Ireland also found the remains of a Dipper in a Peregrine nest (Perry, 1986) and Tony Cross (pers. comm.) similarly recorded a Dipper ring from a Welsh eyrie. With the increase of the Peregrine population in Wales, many pairs of Dippers now nest in close proximity to Peregrines and are potentially vulnerable. However, we know of several pairs of Dippers that still regularly nest immediately below rock faces with a Peregrine nest.

It is surprising that so eloquent a writer and experienced an observer as John Muir wrote, in 1894, in *Mountains of California* that: 'I never knew him [an American Dipper] to be chased by hawks'. Although we can find few references to predators of American, or White-headed, Rufous-throated or Brown Dippers, it seems probable that all these species fall prey to local raptors, as well as to stream-dwelling mammals.

Owls

Owls have been noted as predators of dippers in Europe and in South America. Baake (1982) referred to a Tawny Owl as a possible predator of a Dipper, and in England and Wales metal BTO rings from Dippers have been found in Tawny Owl pellets. In 1988 in mid-Wales, a pair of Tawny Owls nested on a ledge below a bridge where in the previous year Dippers had nested and roosted. Sergio Lasso (1988) found a part-eaten female White-capped Dipper on its nest on the Rio Guayllabamba system in Ecuador, and he concluded that it had been attacked by an Andean Pigmy Owl, a known predator of birds, which occurred in the locality.

Other Predators of Dippers

Other predators of dippers include water snakes in North America (Steiger, 1940), while in Nevada in the USA, Johnson (1952) identified the partly digested remains of a fledgling American Dipper from the stomach of a Brook Trout. The trout was only 10 inches (25 cm) long but as Johnson said: 'it was capable of the predation by use of its relatively cavernous mouth'. He suggested that mortality of this type could be high on streams where the bird is closely associated with large fish. Elliott and Peck (1980) also recorded an American Dipper as prey of a trout. Possibly large Brown Trout or, more likely, Pike may prey on Dippers in Britain and Europe; there is no evidence of this that we can find, although Pike were suspected in the loss of an adult at one nest.

Writing of the Brown Dipper and eastern race of the (White-throated) Dipper in subalpine Tien Shan in the USSR, Kovshar (1979) noted that the Dipper nests were so well protected by mountain torrents that they were inaccessible to terrestrial and aerial predators. He knew of no nest of either species being predated, except by man. Stepanyan (1962), however, found that nests of Brown Dippers in central Tien Shan were better protected than those of (White-throated) Dippers which were not uncommonly entered by small mammals and their contents destroyed. No details were given though of the species of small mammal. It would seem likely that further observations would show that nests of both species are vulnerable to predation, although some nest sites, such as those behind waterfalls, are certainly much safer from predators than others.

Cuckoos and Dippers

Although cuckoos are not strictly predators of nests they do cause loss of eggs and young of host species. In Britain the European Cuckoo very rarely selects a Dipper as host. Only one nest record card from more than 7000 cards for

Dippers submitted to the BTO up to 1990 noted a Cuckoo nestling. This nest, found by Stephen Richardson in Durham, was in a ledge about 1 m into a drainage tunnel beneath a road — an extraordinary site for a Cuckoo to first find and then in which to lay its own egg.

In the Bringhi Valley in Kashmir, Rauf Ali (1967) observed a Brown Dipper being followed by a Cuckoo, which the Dipper then fed. Another instance of a Brown Dipper as a Cuckoo host is recorded by Stuart Baker (1942), who collected a Cuckoo egg (*C. canorus bakeri*) from a nest in Kashmir. Ali and Ripley (1983) noted that the eastern Himalayan race of Brown Dipper was not uncommonly used as a Cuckoo host.

PARASITES

Internal Parasites

Little work has been carried out on internal parasites of dippers, although the blood or bone marrow of three species has been examined for avian blood parasites or haematozoa. None were found in those few Brown Dippers that were checked but in both the American and White-throated Dippers, blood parasites have been recorded. Stabler, Holt and Kitzmiller (1966) and Stabler and Kitzmiller (1970) recorded trypanosomes from American Dippers. In the White-throated Dipper, *Trypanosoma*, *Leucocytozoon* and *Microfilaria* have been recorded (Bennett *et al.*, 1982). Two families of Dipterans, the hippoboscid flies and black-flies or simuliids, have been implicated as vectors (transmitters) of these blood parasites.

In 1989, a small sample of blood was taken from about 20 Dippers in mid-Wales, though these proved negative.

Macey and Strong (1967) reported a new species of trematode or liver fluke from the small intestine of American Dippers. Of 13 birds that they examined from the lower Columbia River in Oregon, 10 were found to be infected with no fewer than 441 trematodes. One from near Washington had three such trematodes but none were found in six birds collected from the Nehalem River in Oregon. The trematode differed from seven other known species of the genus *Laterotrema*, and Macey and Strong named it *L. cascadensis* (*Stromylotrematidae*). This was the first trematode to be found in dippers in the Pacific northwest. One other species of the genus *L. cincli* was found by Yamaguti (1939) in Japan in a Brown Dipper.

Ectoparasites

Ectoparasites, living on the skin or feathers of Dippers, are apparently uncommon judging by the few records of fleas, lice, mites or louse-flies.

Casual inspection of over 3000 Dippers in Wales showed that only one bird in 100–200 had any louse-flies, the large flattened often wingless flies more commonly found on swifts and martins. We have collected two different species of louse-fly from Welsh Dippers: *Ornithomya avicularia* (Linn.) and *O. chloropus* (Bergroth), previously known as *O. lagopodis* (Sharp). These two species of louse-fly have been recorded on a wide variety of bird species (see Rothschild and Clay, 1952; Mawson, 1979) but *O. chloropus* had not previously been found on a Dipper host. *O. avicularia* had been found only once before from a Dipper, in Germany where of 232 Dippers examined by Spitznagel (1985a), just 8 were infested, each with one fly; these were only recorded between July and October.

Since 1985 occasional broods of Dippers have been found in Wales which have been infested with large numbers of Northern Fowl Mites *Ornithonyssus sylviarum* (Canestrini and Fanzago). These are specialised blood-sucking ecto-parasites which commonly infest poultry, pigeons and many other birds, in some instances transmitting viruses. Although they caused apparent distress to the Dipper nestlings (e.g. nestlings unusually calling continuously), the young birds fledged successfully, and one nest was then re-used for a second clutch. Heavily infested poultry can also suffer irritation, weight loss, reduced egg production and even death if loss of blood is great.

In Germany, Spitznagel (1985a) has also found small numbers of these mites on a few Dippers (5 out of 232 examined). The mites, 1–43 individuals on any one bird, were on feathers of the back, breast and belly, and also in the birds' nostrils and in their nests, but only between November and February. Schmid (1985c), too, has described intensive attacks on Dipper nestlings by this mite. In 1984 he found 13 of 45 broods to be parasitised by the mite; this followed four years of low infestation and occurred almost exclusively in broods in nest-boxes constructed of 'eternit', an asbestos and cement mixture.

In California, Halstead (1988) recorded the first instance of American Dippers affected by the Northern Fowl Mite. In 1984 he found a single nestling and its nest infested with approximately 400 *O. sylviarum*. Both nymphal and adult stages were collected from the feather and skin of the nestling, but most mites were in the nest.

Some 54 species of mites were found in seven Dipper nests from our Welsh study area (Fain, Greenwood and MacFarlane, 1991). These came from 46 genera and included pests of stored foods, species able to invade houses or to produce respiratory allergies in man and species commonly occurring in moss, leaf-litter, grassland or dead wood. Rothschild and Clay (1952) record a group of red mites which suck blood from the nasal cavity of many species of bird including the Dipper.

We have found feather lice (Mallophaga) in Dipper faecal samples, probably as a result of birds preening their plumage and swallowing the lice. In 1987 two adult Dippers caught at roost in mid-Wales had lice on their throat feathers. These were subsequently identified by Dr Jim Fowler as *Philopterus cincli.*

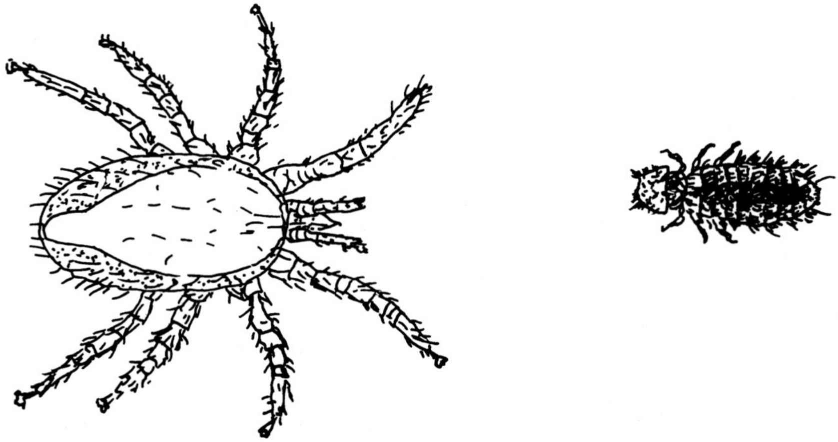

*Parasites: feather mite (*Ornithonyssus sylviarum*) and feather louse*

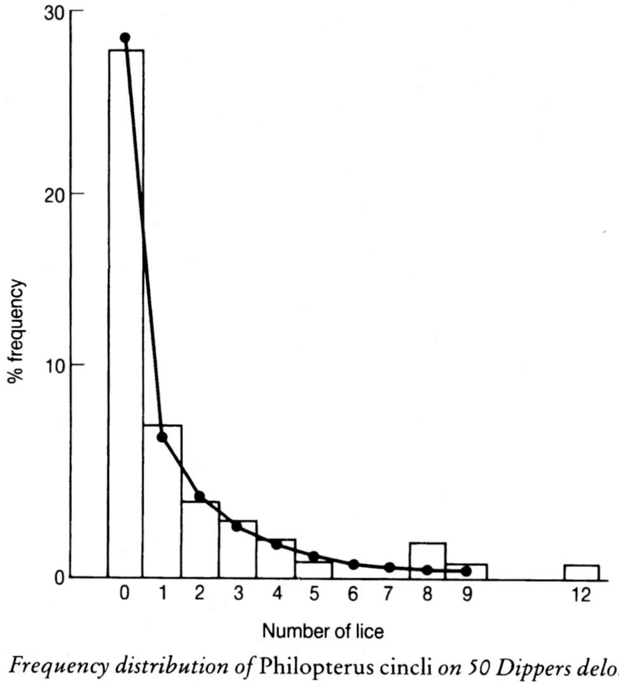

Figure 65. Frequency distribution of Philopterus cincli *on 50 Dippers deloused in central Wales.*

NOTES: Joined closed circles are the expected frequencies of a negative binomial distribution based on sample data (Fowler and Hodson, 1991).

During another night in September 1987 Dippers caught at roost under bridges in mid-Wales were deloused in Fair Isle apparatus (Fowler and Hodson, 1991). Of the 50 birds examined, 27 (54%) yielded 93 lice (mean 1.86 lice per bird, range 0–17). *P. cincli* was the most frequent species (Fig. 65) but a second species, *Myrsidea franciscoloi*, was found on six birds. There was no significant difference in infestation between age or sex classes.

Spitznagel (1985b) also found these two species of feather lice on 43% of 160 full-grown Dippers in his study area in Germany and on 12% of 40 nestlings. *P. cincli* was recorded on birds throughout the year whereas *M. franciscoloi* was recorded only from December to February. The heaviest infestation was after moult in the early autumn and in the months preceding the breeding season. Most feather lice were found in the head region of the Dippers. More males than females (90 : 70) were infested but there was no significant difference.

Four broods of American Dippers in Colorado were found to be heavily infested with feather lice but all fledged normal young (Price and Bock, 1983).

Feather lice are generally quite host-specific and also food-specific. They feed on feathers of a particular part of the host's body and will not survive on other parts (Ash, 1960). A normal louse population on a bird probably has little effect on its host. By preening, and perhaps by bathing, a healthy bird can keep the lice numbers in check. It is believed that only unhealthy birds become very heavily infested.

Many birds harbour fleas and those species which have nests in holes in banks or trees, for example Sand Martins and Blue Tits, often have large numbers of fleas. Dipper nests are similarly often in holes or dark enclosed sites; nests may be re-used for successive broods in a season or even in successive seasons, and birds may use the old nests for roosting in the autumn and winter. The cavity nests and normal domed nests therefore, provide ideal conditions for fleas to breed and to spread from one bird to another. Rothschild and Clay (1952) recorded at least four instances of Sand Martin fleas *Ceratophyllus styx* on Dippers from widely separated areas. This flea is usually strictly host-specific.

We have rarely noticed fleas on Dippers that we have handled, but have not specifically searched for them. Examination of Welsh nests collected at the end of the breeding season has revealed that many harbour two common species of flea: *Ceratophyllus gallinae* and *Dasyphyllus gallinulae* (Malcolm Greenwood, pers. comm.). The former species is generally predominant but *D. gallinulae* is apparently more common in those nests from wetter sites. *C. gallinae*. the so-called hen flea, has been recorded on over 65 different host species in Britain, but is mainly a nest flea. *D. gallinae*, the Moorhen flea, which has also been recorded on about 60 species of bird, occurs more often on the body of the bird. The hen flea apparently prefers elevated nests, or nests in tree tops, i.e. dry nests, whereas the Moorhen flea prefers lower nests in scrub and brambles, i.e. damper sites. This conforms with the preference shown by the latter species for nests of Dippers from wetter sites.

Surprisingly, Breitenmoser-Wursten (1988) reported that attacks by fleas *D.*

gallinulae were a main cause of failures of first broods of Dipper in her study area in the northern Swiss Alps. She noted that 21% of these failures were attributable to fleas with only rising water levels causing greater losses.

Sergio Lasso (1988), who studied a small number of pairs of White-capped Dippers along the river Guayllabamba in Ecuador, found a total of 592 fleas in the nest of a second brood.

Blow-fly larvae of the genus *Protocalliphora* also suck blood from the nestlings of many species of nidicolous birds. It is no surprise that dippers may also be victims of blow-flies. From 1984 to 1987 Halstead (1988) working along creeks in Fresno County, California, found 11 of 45 nests of the American Dipper to be parasitised by *P. aenea* or *P. braueri*. Up to six blow-fly larvae were embedded subcutaneously on the head, wings and legs of the dipper nestlings. Between 10 and 40 puparia were found in some nests. Halstead also noted some previously unpublished records of *P. aenea* parasitising American Dippers in Utah and Colorado.

Blow-fly larvae can cause nestling mortality in tits or, through the blood they take, contribute to nestling stress and perhaps cause reduced post-fledging survival (Gold and Dahlsten, 1983). Halstead only attributed mortality of one dipper nestling at 14 days old to *P. braueri* parasitisation. This nestling was infested with two larvae in a wing and five in the head; these caused limited movement of the mandible and probably impaired hearing and vision. Other parasitised nestlings did not appear to be seriously impeded.

Clearly there is much work still to be carried out on the internal and external parasites of dippers, and on the effect of any parasites on the birds. The low incidence recorded may be due to lack of observations or to the solitary nature and rather isolated lifestyle of dippers, but it may be due to the time that dippers spend in or underwater or to the dippers' frequent and careful preening of their plumage. This is, of course, necessary to maintain 'waterproofing' of the feathers. Dust-bathing and the 'anting' behaviour shown by many species of bird are often associated with feather care and anti-parasite measures. Perry (1986), in this context, reports an interesting piece of behaviour observed by A. Johnston. A Dipper in Ireland caught a small Brown Trout and then rubbed the fish into its breast feathers, flanks and wings. A few days later a Dipper was seen catching another fish which it also then rubbed into the outer wing feathers and under the wing feathers before swallowing it.

True anting by Dippers has been recorded in Germany (Creutz, 1952; Heinroth and Heinroth, cited in Armstrong, 1955), and in England, both in Gloucestershire (Angus, 1980; Mary Muller, pers. comm.) and in Devon (Roskell, 1982). Creutz describes a female Dipper climbing to a turf-covered rock, picking up ants and placing these on and under the wing with poking and rubbing movements. Roskell watched a Dipper picking up ants (*Lasius niger*) from a stone wall at the edge of grass, then preening, at the same time bobbing or 'dipping' vigorously. Ants found in the faeces of Norwegian Dippers by

Ormerod, Efteland and Gabrielsen (1987) may have been swallowed following anting behaviour. Goodge (1959) also observed possible anting behaviour in the American Dipper.

Stream in hills with conifers on catchment

CHAPTER 10

The impact of humans on dippers

The activities of humans have had some influences on the range and status of all five species of dipper. Pollution, major river engineering works and over-abstraction of water for agriculture and domestic and industrial supplies, as well as direct persecution, have adversely affected the distribution and numbers of dippers. However, the construction of walls, bridges and other buildings near rivers has provided safe nest and roost sites, and in some areas allowed dippers to occur in higher numbers than might otherwise be the case.

PERSECUTION BY HUMANS

Direct persecution of dippers, either for food or because of some fear that the birds might affect fish populations, may have had local impacts. There is little evidence today that dippers are widely hunted, but in southern Europe and North Africa dippers still fall prey to humans. The Dipper population on Cyprus became extinct in 1945 (Flint and Stewart, 1983). The reasons for this are unclear but Cyprus, along with Malta, has one of the worst records for the slaughter of its small birds, although is also very dry and eroded. In North Africa Dippers are still hunted and when we were carrying out fieldwork in 1986 in the High Atlas Mountains in Morocco, the Berbers requested that we give them the birds that we had mist-netted. We later learnt from Alistair Scott, on the word of a local guide, that the Berbers regarded Dippers as having aphrodisiacal properties.

Until the beginning of the twentieth century a bounty was paid in parts of Germany and Scotland for slain individuals or the destruction of Dipper nests. Those killing Dippers on some Scottish estates had as an incentive, special rights to fish salmon with rods during the close season. Elsewhere, as on the Duchess of Sutherland's estate, a lure of 6*d*. a corpse encouraged the slaughter of 548 'king's fishers' or water ousels between March 1831 and March 1834. A further 368 water ousels appeared on the lists of exterminated vermin in Reay county, also in Sutherland. We know today that this carnage was misguided and that Dippers are very unlikely to have any adverse effect on fish stocks.

In 1952 Jones and King, who observed a Dipper visiting a salmon observation tank, found that the bird readily ate eggs. They commented though that as the bird could only reach stray eggs which would never have the chance to develop, the Dipper could not be regarded as a serious predator of salmon. Dietary studies (see Chapter 4) have also shown that few prey items are fish, most of these being Bullheads rather than small salmon or trout. At most a pair of Dippers resident all year and rearing two broods could only crop 2–5% of the annual salmonid production within their territory. The natural mortality amongst the young stages of salmonids is very high anyway, and many other factors such as overfishing at sea, the use of monofilament nets and pollution of rivers and estuaries cause the real damage to fish stocks.

Despite this evidence, Dippers in a few areas in Scotland continue to be shot and trapped (Watson, 1972), or their nests knocked down (anon., pers. comm.). Four recoveries of ringed Dippers in Britain up to 1985 were of shot birds.

The American Dipper has also been subject to some persecution. Ehinger (1930) remarked on the accusations brought against this species for eating fish fry, salmon, trout and other spawn. He defended the birds though, and stressed 'the prodigious waste of eggs and fishes' through surplus production. He concluded that American Dippers had a negligible effect. Bent (1948) also noted that the bird's habit of eating spawn and fry of salmon and trout, had made many enemies for it among sportsmen and especially among managers of fish farms. Bent likewise did not believe that American Dippers had any serious impact on fish stocks under natural conditions, but admitted that in the artificial situation at a fish farm, they may do damage. A study in 1924 by J. A. Munro (quoted by Bent) into the relation of this species of dipper to fishing interests in British Columbia and Alberta, mentions the very different attitudes prevailing amongst managers of hatcheries at that time. For example, one tolerant manager believed that any damage an American Dipper may do through eating salmon eggs and fry would be balanced by eating eggs and fry of the salmons' competitors and predators. At the other extreme was a manager who claimed that in one winter some 10,000 Cut-throat Trout fry were taken from his pond by American Dippers. Munro concluded from his study that the consumption of spawn was not considered serious by fishery officials, but that when the artificially propagated fry had been placed in retaining ponds, then birds could take significant numbers of young fish. Because other birds soon moved in to replace those which were shot, Munro wrote: 'The obvious remedy is to screen the surface of retaining ponds with fine mesh wire netting. This will adequately protect the fry and render it unnecessary to destroy a songbird of high aesthetic value.'

That was written in 1924. This lesson is worth restating today, especially with the proliferation of fish farms in Scotland and Wales in recent years, and furthur conflict between man and fish-eating birds.

Humans also collect dipper eggs although this is less of a threat today than formerly. Shaw (1978) in his analysis of nest failures found that the main recorded reason for loss of Dipper nests was human interference, with removal of eggs accounting for 30.4% of nest failures. There may be a bias though towards reporting nests which are easily accessible to the observer and hence to a human predator. In our study area in mid- and south Wales we estimate that at most, 10% of nest losses are due to people removing eggs.

HABITAT MODIFICATION

Humans have for centuries inadvertently affected all five species of dipper by altering the habitats in and around rivers. Dam schemes, pollution, erosion, drainage schemes and the creation of hazards and obstacles are all important in this respect.

Hazards and Obstacles

Dippers are occasionally killed through colliding with man-made structures or through becoming entangled in wire or nylon line. Robson (1969) reported a Dipper impaled on barbed wire. Juveniles are particularly prone to accidents and causes of ringing recoveries include such varied reasons as flying into glass or into fences, falling down chimneys, drowning in a reservoir of oil, drowning ironically, through becoming entangled in an anti-predator net at a fish farm and one unfortunate bird flying into an extractor fan. Others (eight recoveries up to 1985) are killed by vehicles.

The litter left by a few anglers, particularly discarded nylon line and hooks, or line entangled in trees and bushes, is an unnecessary hazard. Dippers and other waterbirds can and do become caught up in this and suffer lingering deaths.

When Dippers collect moss as nest material from stones in the river bed, they may accidentally pick up and incorporate bits of nylon line into the nest. At one site in the Wye Valley (Welsh borders) a female was found dead, hanging suspended from her nest by such a piece of line. Subsequently, line was found in 5 of 14 Grey Wagtail and Dipper nests in the valley.

Recreation Pressures

Increased recreation brings its own problems for birds, especially during the breeding season. Canoeing and boating may cause some disturbance but there is little evidence of any great problem so far as Dippers are concerned. Unfortunately, innocent groups of people picnicking by, or paddling and bathing in, rivers, especially by nest sites, can cause nest failures. Through remaining too long too close to the nest, they prevent the adult birds from returning. Anglers can cause similar disturbance. Shaw (1978) attributed 25% of nest failures to desertion, some of these due to Dippers being kept off their nests by the presence of anglers or picnic parties. Eggs or small young become chilled under these circumstances. Although the clutch or brood may be lost, Dippers may relay if the season is not too far advanced. Fortunately, because Dippers lay early in the year, most in Britain having young by mid-April, first broods will usually have fledged before the height of the picnicking season. It cannot be claimed that recreation activities have any known significant detrimental effect on Dipper populations, but they can jeopardise the breeding success of individual pairs, and be a considerable annoyance to anyone carrying out a field study!

Destruction of Nest Sites

Man-made structures such as walls and bridges, have provided Dippers with relatively safe nest and roost sites. Bridges are often preferred to natural sites even where rock ledges and crevices exist, and on lower stretches of rivers they may provide the only suitable nest sites. It is unfortunate that the old stone bridges or girder bridges, which are ideal for Dippers, are not able to support modern traffic such as heavy milk tankers or lorries laden with timber. Consequently they are gradually being replaced, usually by concrete box bridges. Unless these have drainage holes or ledges, they are of little use to dippers, other birds or even bats, and many old traditional nest and roost sites are being lost in this way throughout the Dipper's range in Britain and Europe.

Nest-boxes below bridges, or incorporated into the bridge sections at the time of their construction, provide one solution. In Britain, Lewis and Williams (1984) described a design for a Dipper nest-box for use by river managers. Highway authorities of some county councils in Wales and England have also been urged to consider erecting boxes under bridges for which they are responsible. The National Rivers Authority in parts of Wales, and many local Dipper enthusiasts, have started nest-box schemes. In the Grampians, Rae and Hardey (1979) put up seven boxes on stretches of river where there were no apparent nest sites. All seven were occupied by Dippers the first season after they were in place. In parts of Europe, Dipper boxes have also been used to try to sustain or increase breeding populations. In Hungary, Horvath (1987) has tried to halt the decline of Dippers by providing nest-boxes on rivers where bridge nest sites have been destroyed. Wagner (1985) in 1984 initiated a scheme in Austria to encourage road and bridge inspectors to put up nest-boxes under new bridges, whilst in Germany nest-boxes have been widely used to sustain breeding populations of Dippers. In his study area, Kaiser (1985) found 64% of Dipper nests were in nest-boxes that had been erected to try to increase the very small population on rivers. Rockenbauch (1985a) also drew attention to the need to retain old buildings, mills and especially bridges by rivers, or to create new niches and use nest-boxes, to maintain breeding densities at existing levels. He found that more than a third of Dipper broods recorded on 420 km of river between 1956 and 1984 were in nest-boxes. Nest-boxes have also been used successfully in North America.

Reservoirs and Hydroelectric Schemes

The regulation of rivers by humans through the construction of dams, reservoirs and hydroelectric schemes has had a profound effect on the ecology of rivers throughout the world. Releases of water may raise the water level in the river, wash out nests and prevent birds from feeding. Channels may be scoured and eroded below the dam. New concrete channels may be constructed which are poor habitats by comparison with the original watercourses with their diverse habitats of shoals, riffles, margins and deeper pools. Water from a deep-water reservoir if drawn off from the lower layers may be cold, low in oxygen and have a high metal content. These factors may all adversely affect the invertebrate and fish populations on which certain birds and mammals depend.

Reservoirs also eliminate the river habitat which they flood and replace. Shooter (1970) recorded a Dipper nest site in the Goyt valley in Derbyshire lost by the building of the Erwood reservoir, and noted that in the past, 24 km of river had been flooded by reservoir construction in the gritstone area of the county, probably destroying 6–9 territories. One of the most worrying aspects of reservoir construction and irrigation schemes in arid areas such as around the Mediterranean, is that flows in the rivers are often considerably reduced, so much so that they run dry for much of the year. The tourist industry and farmers demand an ever-increasing supply of water for swimming pools or for their 'thirsty' crops such as oranges and tomatoes. Detrimental effects on those birds such as Dippers which depend on running water throughout the year have been

Concreted and canalised stream in Sicily; a major threat to Dippers here (M. Sara and G. Sorci)

recorded in Sicily (Maurizio Sara and Gabriele Sorci, pers. comm.). Sara and Sorci also attribute declines in the Dipper populations in Sicily to habitat destruction caused by channelisation and the construction of reservoirs, aqueducts, pipelines and wells. They believe the remnant isolated populations are on the verge of extinction.

Other species have been adversely affected by such developments. For example, in north-west Argentina, reservoir construction and associated holiday villages, hydroelectric schemes and irrigation schemes to provide water for tobacco and sugar cane crops, have locally had an impact on the scarce Rufous-throated Dipper and possibly other river birds. Admittedly some wetland birds, including herons, ducks, coot, ibises and spoonbills have benefited from new reservoirs and expanded their ranges.

Wagner (1985) reported that 12 of 73 Dipper territories in Carinthia in Lower Austria were destroyed by the construction of water-power stations and modern bridges; four more power stations were to be erected on 40 km of the R. Drau. He proposed a minimum water quantity of 0.5 m^3/s at diversion pipes to water-power stations and improved design of channels to make them more 'natural' and acceptable to Dippers. Horvath (1987) suggested that river regulation, in destroying Dipper breeding and feeding sites, partly explained the worrying decline of the species in Hungary.

Power stations may sometimes have a beneficial effect on dippers too. Price and Bock (1983) found that a hydroelectric plant on Boulder Creek in Colorado kept much of the stream free of ice in the winter, so providing additional feeding habitat for American Dippers in this critical period. The plant did, however, discharge water at peak power demand periods which could raise water levels by 0.5 m. This increase in water levels could make it more difficult for birds to feed at these times.

Dam across a mountain river in the Sierra de Aconquija in Argentina

Deforestation

A major effect on the whole ecology of rivers is caused by the removal of natural forest from the watersheds. Forests help to maintain a steady flow of water into streams through regulating run-off from the soil. Once the forest cover is removed, the bare soil may be compacted or eroded and rain runs quickly over the surface and down the slopes. This results in flash floods after rain, but drought for the rest of the year.

We may be seeing the results of removal of forest cover in the Himalaya in the increasing number of serious floods in Bangladesh although some studies do suggest otherwise. We are only just beginning to look at the impact of such changes on the river itself and on all that the river supports. The Brown Dipper may already have been affected over much of its range in the lower foothills, where cultivation has replaced the Sal forests and other tree cover. Increased sedimentation in these rivers, bank erosion, especially when the fringing Himalayan alders are cleared, and spatiness all affect those invertebrates on which the Brown Dippers depend. During a survey of rivers in the Middle Hills of Nepal in 1991, we found that Brown Dippers were absent from streams where flows were greatly reduced by diversion of water into irrigation channels feeding rice paddies, and with heavy deposits of sediment on their beds. These same streams supported few caddis larvae or mayfly nymphs by comparison with those with a good flow. Exactly the same may apply in those parts of the White-capped Dipper's range in the Andes where tropical forests have been cleared. Deforestation in Argentina is considered a threat to the Rufous-throated Dipper by R. Vides Almonacid (in Collar *et al.*, 1992). Loss of deciduous tree cover and land degradation in upland Britain may also in the past have resulted in lower densities of breeding Dippers than existed formerly.

Pollution

Humans have long used rivers to 'export' wastes but during the last two centuries, the range and toxicity of wastes discharged into rivers has considerably increased. In 1924 Kathleen Carpenter (1924) studied the pollution of the Rivers Ystwyth, Rheidol, Melindwr and Clarach in north and west Wales, where as a result of drainage from old lead, zinc and silver mines, invertebrate and fish faunas in many reaches were restricted. J. H. Salter, as long ago as 1895, commented that Dippers were absent from the Ystwyth and other rivers 'fouled by lead washings'. Mining on the Ystwyth ceased in 1921 and there has since been much chemical and biological improvement. By 1984 Dippers were breeding on parts of the Ystwyth, although they were still absent from the relatively acidic and unproductive waters further upstream.

In industrial areas of south Wales, pollution over the last two centuries by sewage and industrial wastes led by 1971 to about 60% of the rivers draining the coalfield being described as 'grossly polluted' or of 'doubtful quality'. The consequences of such pollution is well known; generally, groups such as stonefly and mayfly nymphs, and caddis larvae, which form an important part of the diet of breeding Dippers, are most affected by polluting discharges. Dippers therefore became scarce on these south Wales rivers although with recent trends in improvement of water quality of rivers in former industrial areas, they are now returning to breed on tributaries of the Taff, Ebbw and Sirhowy. Some rivers, however, still have problems caused by iron-rich water flowing from abandoned coal mines and forming an ochrous deposit on the river bed, which eliminates many invertebrates and fish. There are over 90 ferruginous discharges in Wales with at least 63 km of otherwise suitable Dipper river in Wales affected in this way; some 17 km of the R. Pelenna in west Glamorgan, the so-called 'Orange River', is polluted by iron deposits. In the north of England there are more than 160 significant discharges from old mines, with similar problems in the area covered by the Forth and Clyde River Purification Boards.

In 1975 about 1500 km of river in those areas of England and Wales where Dippers breed, were classified as being of poor quality or grossly polluted, with a further 1800 km of doubtful quality. An estimate of five pairs of Dippers per 10 km would mean that those polluted waters represent habitat for 1650 pairs of breeding Dippers, probably 5–6% of the current population. Whilst not all the polluted rivers may have been physiographically or otherwise suitable for breeding Dippers, these calculations at least provide an indication of pre-industrial populations.

There is now much tighter control on chemical pollution by factories but deliberate spills or discharges do occur, as well as inevitable accidents. Tankers may discharge their waste loads into a convenient watercourse rather than take them to an official dump. An acute pollution incident can undo years of work in improving water quality and encouraging fisheries. If they occur in the breeding season, such incidents can cause catastrophic effect on those birds nesting downstream of a spill or discharge.

Rivers in Europe and elsewhere in the world have also suffered from chemical and metal pollution. Santamarina (1988) found that Dippers were absent on rivers in Galicia, northern Spain, which had been polluted with copper, although these rivers do drain granite and receive acid deposition too. Lachenmayer, Kunze and Holzinger (1985) detected eight metals (notably copper, aluminium, lead, zinc and nickel) in invertebrates and in eggs of Dippers and Grey Wagtails from a contaminated stream in western Germany, although they did not report any adverse effect on breeding performance as a consequence. Mercury levels in freshwaters have declined since the 1960s, but mercury was found in Dipper eggs from Wales, Scotland and Ireland from 1988 to 1991, although there was much annual variability. The highest levels were recorded in eggs from south-west Ireland in 1990 but even here concentrations were not high enough to be problematical (Ormerod and Tyler, 1992).

On the R. Marscheid in western Germany, Monig (1985) found that a number of first clutches of Dippers failed in successive years. Eggs from these clutches contained levels of polychlorinated biphenyls (PCBs) at concentrations which Monig believed had affected the birds' fertility. Highest concentrations were in those nests from the headwaters and decreased downstream. We have also found levels of PCBs in some Dipper eggs in Wales, Scotland and Ireland as high as, or higher than, those found in raptors in Britain and in insectivorous waterbirds elsewhere, but levels between 1988 and 1993 have fluctuated widely (Ormerod and Tyler, 1990b; 1992; 1993a). PCBs, once widely used in a range of substances such as lubricants, adhesives, paints, sealants and plastics, were banned in 1986, but these highly stable compounds now occur throughout the world's freshwaters and oceans, and are known to accumulate in the fat and liver of mammals, inducing stress, disease and infertility (Mason, 1991). Whilst we have no evidence that PCBs have affected Dipper populations, it is of interest that one PCB congener occurred in significantly higher concentrations in eggs of Dippers that failed to develop than in fertile eggs (Ormerod and Tyler, 1992). Some contribution to egg failure therefore cannot be ruled out although only a small proportion of eggs fail. Eggs from certain streams consistently contain higher levels of PCBs than in those collected from other streams; in mid-Wales we found from 1990 to 1993, higher levels of PCBs below a certain point on the Afon Mule, a tributary of the R. Severn, than on the upper Mule or rivers elsewhere in Wales. The source of the contamination was traced to the leachate running into the stream from an old waste tip on which transformers had been disposed. This again indicates the value of using eggs of Dippers for monitoring contamination.

In Ecuador, Sergio Lasso (pers. comm.) reports that streams draining the industrial areas of Quito no longer support White-capped Dippers. In Argentina, M. Nores (in a letter to W. Belton in 1985) suggested that as Rufous-throated Dippers occurred on a river in the Aconquija Mountains in Tucuman province just below a reservoir and a resort, they could tolerate some degree of pollution. However, King (1978–79) considered that grossly polluted streams were a threat, and on a visit to the Aconquija range in October 1993 we found that the river referred to by Manuel Nores was now grossly polluted (Tyler, 1994). Enrichment with nitrates or phosphates had caused excessive filamentous algal growth all down the river, with blankets of algae covering rocks and shoals. We recorded only one Rufous-throated Dipper on a 7-km section below the dam and failed to find any below this section, although on a clean tributary immediately above its confluence with the main river, there were two pairs in about 1 km. It is of interest that nests of both South American species sometimes include plastic and refuse in their linings, perhaps an indication of the habits of the human population close to accessible rivers and bridges!

The contamination of streams by farm wastes, especially the highly toxic silage and slurry, is commonplace in dairying areas of south-west England and Wales.

Chronic pollution of small order streams, or acute pollution from a leaking silo or accidental spill kills aquatic invertebrates and fish. Results of a survey in 1989 by the RSPB of streams in Devon and Dyfed showed Dippers to be scarcer on those watercourses polluted by farm waste and with impoverished invertebrate faunas, than on cleaner streams with more abundant caddis larvae and mayfly nymphs (Tyler, 1990). Results of a study on the R. Wye in Derbyshire also suggested that the number of Dippers was reduced by organic pollution (Cole, 1990).

Some rivers are adversely affected by nitrate and phosphate enrichment from fertiliser run-off and sewage, but this is less of a problem in the north and west of Britain where most Dippers occur, than in the lowlands. The Argentinian river suffering from massive eutrophication, as noted above, suggests that nutrient enrichment may become a wider problem for river birds in those countries where rivers are treated as drains for waste disposal.

Persistent organochlorine pesticides such as DDT, dieldrin and endrin, were once widely used in agriculture in seed dressings, sprays and sheep dips, and although now banned, traces still occur in rivers. Other sources of contamination include timber treatment works and carpet factories. Generally, concentrations of organochlorines in livers of piscivorous waterbirds has declined during the 1970s and 1980s (Newton, 1991). Detectable levels of these chemicals were found in some deserted and infertile Dipper eggs collected in Wales from 1988 to 1993, with the highest levels in eggs from Scotland in 1990 (Ormerod and Tyler, 1990b; 1992; 1993a). There were differences between catchments in Wales with levels of dieldrin highest in those areas where sheep densities were highest, suggesting a legacy from use in sheep dips (Ormerod and Tyler, 1993a). Much higher levels must have occurred in the past when dieldrin was widely used in dips, but if Dipper populations were reduced, this has gone undetected.

It is ironic that as water quality in formerly polluted rivers in industrial areas improves, a relatively new threat has appeared in apparently pristine areas of upland Britain, Scandinavia, central Europe and Canada. In these countries areas overlying base-poor rocks, as in west and north Wales and much of Scotland, are vulnerable to acidification. Emissions of sulphur dioxide and nitrogen oxides from power stations, vehicles, agricultural practices and other sources have greatly increased in Britain since the start of the Industrial Period. Initially effects were limited to areas immediately around the source of pollution. For instance, vegetation on moors in the Peak District close to Manchester were adversely affected in the early part of the century. Pollutants are now carried further afield from high chimney stacks and it is now known that sulphur emissions in Britain have contributed to the acidification of thousands of lakes and rivers in southern Sweden and Norway, as well as to the acidification of many lakes and streams in upland Wales, England and Scotland. For example, the National Rivers Authority estimates that about 12,000 km of watercourse in Wales have already been affected by increased acidity.

River Irfon in mid-Wales: an acidified stream with conifers in catchment and a very low Dipper population

The planting of Sitka and Norway Spruce and other conifers in upland acid-sensitive areas in Britain has exacerbated the effects of acidification. Their needles provide a large surface area for trapping air-borne sulphates and other pollutants, particularly from mist and clouds. Following rain, the sulphur particles are converted to weak sulphuric acid which falls to the forest floor, into drainage channels and into the streams. An additional problem is that aluminium, normally present in the soil in an insoluble form, is mobilised by acidic water, and enters the watercourses to the detriment of fish and some invertebrates. The headwaters of rivers, often important spawning areas for fish, are the first to suffer. Trout and salmon fry are very susceptible to acidic waters with a high aluminium level; they die and gradually fish populations decline. Some invertebrate groups, notably mayflies and caddis flies, are also intolerant of acidic waters. Their nymphs and larvae, which form an important part of the Dipper's diet particularly in the breeding season, are consequently scarce or absent in the acidified upland streams. Calcium, essential for egg-shell formation and growth of nestlings, is in short supply on these streams. We know that as a result of low calcium and limited food, few pairs of Dippers breed on acidic streams by comparison with better quality streams; Dippers on acidic streams also have longer territories; they lay later, lay fewer eggs and rear fewer young than elsewhere (Ormerod and Tyler, 1987b; Ormerod *et al.*, 1991). Moreover, they spend more time foraging, swimming and diving and less time resting than birds on better quality streams. Consequently, birds at acidic sites will have less time for such activities as preening and watching out for predators (O'Halloran *et al.*, 1990).

The adverse effect of acidification on Dippers has been reported from other areas in Britain such as in Galloway by Vickery (1991), from Germany where Kaiser (1985) reported reduced numbers of birds on acidic brooks and a reduction in breeding success, and in Belgium (Daulne 1990). There is also anecdotal evidence from Scandinavia (Eriksson, pers. comm.). Sulphur dioxide pollution in parts of eastern Europe is so severe that damage to trees has occurred. Pawl Mielczarek reported (per Kevin Standring) that in south-west Poland, where there is extensive tree damage downwind of lignite coal-fired power stations on the Polish/Czechoslovakian/east German borders, Dippers are apparently absent.

Loss of fish as well as of Dipper populations is of concern to statutory river authorities (the National River Authority in England and Wales). Government commitments to bringing about a reduction in sulphur emissions are welcomed but to prevent further damage and even allow acidic streams to recover, it is necessary to bring about a much greater reduction in sulphur emissions (90%) throughout Europe, and hence much reduced deposition levels, than the UK Government is willing to commit itself to. Guidelines issued by the Forestry Commission suggest possible measures to combat the exacerbating effects of conifers; these include keeping trees well clear of streamsides and no longer having drainage channels from the forests discharging directly into watercourses. There is as yet little evidence that such practices have any ameliorating effect on acidification, and probably they have none, although they may be beneficial for other reasons. Another possible remedy is to lime acidified lakes or rivers, or to spread lime on upland catchments, to neutralise the acidity. This can, however, adversely affect other components of the upland environment. Thus lime spread on to bogs and pools kills the *Sphagnum* bogmoss and other calcifuges, and may also adversely affect some invertebrate groups. These areas are vital for upland birds in the breeding season. Chicks of Red Grouse and Golden Plover, for example, feed on insects in these very areas. Liming is also very costly and the lime must be obtained from some source with the consequent environmental impacts of quarrying. The appropriate rate and amount of lime application on to a catchment is still unknown. Liming on many upland catchments may therefore be inappropriate; far better to reduce the pollution at source, and to control further afforestation of vulnerable areas.

Healthy dipper populations on upland rivers throughout the world indicate healthy river ecosystems. If dippers are to survive in their present ranges, we must ensure that rivers retain their flows and that their waters remain unpolluted. Where damage has already occurred due to inappropriate land use or from humans discharging his various waste products of the twentieth century, we can only hope that collectively humans will make the decision to give priority to redressing this damage, and to managing rivers in the future in a more sustainable way. Only then may future generations enjoy the delight of observing these superb birds.

APPENDIX 1

List of Latin names

BIRDS

Bluebird	*Sialia*
Bluebird, Himalayan	*Grandalia coelicolor*
Cinclodes, Bar-winged	*Cinclodes fuscus*
Cinclodes, White-winged	*C. atacamensis*
Crow	*Corvus*
Crow, Carrion	*Corvus corone*
Cuckoo, European	*Cuculus canorus*
Dipper, American	*Cinclus mexicanus*
Dipper, Brown	*C. palassii*
Dipper, Eurasian or White-throated	*C. cinclus*
Dipper, White-capped	*C. leucocephalus*
Dipper, Rufous-throated	*C. schulzi*
Dipper, White-bellied	*C. cinclus leucogaster*
Duck, Blue	*Hymenolaimus malacorhyncos*
Duck, Torrent	*Merganetta armata*
Duck, Harlequin	*Histrionicus histrionicus*
Flycatcher, Spotted	*Muscicapa striata*
Forktail, Spotted	*Enicurus maculatus*
Forktail, Slaty-backed	*E. schistaceus*
Forktail, Black-backed	*E. immaculatus*
Forktail, Little	*E. couleri*
Grebe, Little	*Podiceps ruficollis*
Grouse, Red	*Lagopus lagopus*
Gull	*Larus*
Jackdaw	*Corvus monedula*
Jay	*Garrulus glandarius*
Kestrel	*Falco tinnunculus*
Kingfisher, Common or Eurasian	*Alcedo atthis*
Kingfisher, Belted	*Megaceryle alcyon*
Kingfisher, Half-collared	*Alcedo semitorquata*
Kingfisher, Ringed	*Ceryle torquata*
Magpie	*Pica pica*
Martin, House	*Delichon urbica*
Martin, Sand	*Riparia riparia*
Merlin	*Falco columbarius*
Moorhen	*Gallinula chloropus*

Ousel, Ring	*Turdus torquatus*
Owl, Andean Pigmy	*Glaucidium jardinii*
Owl, Tawny	*Strix aluco*
Peregrine	*Falco peregrinus*
Phoebe, Black	*Sayornis nigricans*
Plover, Golden	*Pluvialis apricaria*
Redstart, Black	*Phoenicurus ochrurus*
Riverine Flycatcher	*Monachella mulleriana*
Robin, American	*Turdus migratorius*
Sandpiper, Common	*Actitis hypoleucos*
Sandpiper, Spotted	*A. macularia*
Shaketail	*Cinclodes*
Sparrowhawk	*Accipiter nisus*
Thrush, Blue Whistling	*Myiophonus caeruleus*
Thrush, Song	*Turdus philomelos*
Thrush, Varied	*Ixoreus naevius*
Tiger-heron, Fasciated	*Tigrisoma fasciatum*
Tit	*Parus* spp.
Tit, Blue	*P. caerulea*
Torrent-lark	*Grallina bruijni*
Tyrannulet, Torrent	*Serpophaga cinerea*
Tyrant, Flycatcher	*Phylloscartes*
Tyrant, Wagtail	*Stigmatura*
Wagtail, Pied	*Motacilla alba yarelli*
Wagtail, White	*M. alba alba*
Wagtail, Grey	*M. cinerea*
Waterthrush, Louisiana	*Seiurus motacilla*
Waterthrush, Northern	*S. noveboracensis*
Water-Redstart, White-capped (River Chat)	*Chaimarrornis leucocephalus*
Water-Redstart, Plumbeous	*Rhyacornis fuliginosus*
Wren (Winter)	*Troglodytes troglodytes*
Wren, House	*T. aedon*

MAMMALS

Beaver, American	*Castor canadensis*
Beaver, European	*C. fiber*
Cat, Domestic	*Felis cattus*
Cat, Wild	*Felis sylvestris*
Desman, Pyrenean	*Galemys pyrenaicus*
Polecat	*Mustela putorius*
Marten, American	*Martes pennanti* and *M. americana*
Mink, European	*Mustela lutreola*
Mink, American	*M.vison*
Otter	*Lutra lutra*
Rat, Brown	*Rattus norvegicus*
Shrew, European Water	*Neomys fodiens*
Shrew, Mediterranean (and Asian) Water	*N.anomalus*
Skunk	*Mephitis mephitis*
Stoat	*Mustela erminea*

Weasel, European	*M.nivalis*
Weasel, American	*M.frenata*
Woodrat, Bushy-tailed	*Neotoma cinerea*

FISH

Bullhead or Miller's Thumb	*Cottus gobio*
Pike	*Esox lucius*
Salmon	*Salmo salar*
Trout, Brook	*Salvelinus fontinalis*
Trout, Brown	*Salmo trutta*
Trout, Cut-throat	*S. clarki*
Trout, Rainbow	*S. gairdneri/Oncorynchus mykiss*

PLANTS

Alder	*Alnus glutinosa*
Alder, Andean	*A. jorallensis*
Alder, Himalayan	*A. nepalensis*
Beech	*Fagus sylvatica*
Birch, Silver or Downy	*Betula pendula* or *B. pubescens*
Bracken	*Pteridium aquilinum*
Bramble	*Rubus* spp.
Cottonwood (Aspen)	*Populus tremula*
Grass, Purple Moor	*Molinia caerulea*
Ivy	*Hedera helix*
Oak	*Quercus* spp.
Pine, Ponderosa	*Pinus ponderosa*
Rush	*Juncus* spp.
Sal	*Shorea robusta*
Sedge	*Carex* spp.
Spruce, Norway	*Picea abies*
Sycamore	*Acer pseudoplatanus*
Willow	*Salix* spp.
Woodrush	*Luzula* spp.

APPENDIX 2

Quadratic regressions ($y = a + bx + b'\sqrt{x}$) relating the probability of recapturing Dippers after fledging in relation to time of hatching.

	a	b	b′	F	100 r^2	P
All sites	−0.27	−0.10	0.38	20.1	93.1	0.02
Acid streams	−0.47	−0.19	0.66	15.0	90.9	0.03
Circumneutral streams	−0.22	−0.08	0.32	17.8	92.3	0.02
All sites:						
Brood size						
5	−0.39	−0.15	0.54	37.5	96.2	0.01
4	−0.22	−0.10	0.36	9.8	83.0	0.03
3	−0.34	−0.14	0.49	4.8	70.0	0.1
2	−1.56	−0.41	1.68	15.23	93.2	0.05

NOTES: Times were divided into the seven categories listed in Table 11. From Ormerod and Tyler (1993c). (See Figure 46 on p. 127.)

References

ALATALO, R. V. & LUNDBERG, A. 1984. Density dependence of breeding success of the Pied Flycatcher (*Ficedula hypoleuca*). *J. Anim. Ecol.* 53, 969–77.

ALDER, J. 1957. The Dipper's winking. *Brit. Birds* 50, 267–9.

ALDER, J. 1963. Behaviour of dippers at the nest during a flood. *Brit. Birds* 56, 73–6.

ALDER, J. 1966.The cheerful dipper. *Birds* 1, 6–9.

ALI, R. 1967. The Brown dipper (*Cinclus pallassi*) as a host of a Cuckoo (*Cuculus* sp.). *J. Bombay Nat. Hist. Soc.* 64, 561.

ALI, S. & RIPLEY, S. D. 1983. *Handbook of the Birds of India and Pakistan.* Compact edition. Oxford University Press.

ANDERSSON, J. S. & WESTER, S. A. L. 1971. Length of wing, bill and tarsus as a character of sex in the Dipper *Cinclus cinclus. Ornis Scand.* 2, 175–9.

ANDERSSON, J. S. & WESTER, S. A. L. 1972. Body weight of wintering Dippers *Cinclus cinclus cinclus* (L). *Ornis Scand.* 3, 39–43.

ANDERSSON, J. S. & WESTER, S. A. L. 1973. Sex and age ratios of wintering Dippers *Cinclus c. cinclus* (L). *Ornis. Scand.* 3, 39–43.

ANDERSSON, J. S. & WESTER, S. A. L. 1975. Studier av stromstare (*Cinclus c. aquaticus*) in Norge 1968–72. *Fauna och Flora* 70, 257–65.

ANGUS, D. D. 1980. Dipper's 'anting'. *J. Glos. Nat. Hist. Soc.* 31, 398.

ARMSTRONG, E. A. 1955. *The Wren.* Collins, London.

ARNDT, H. 1979. Wasseramseln im Winter. *Gef. Welt.* 103, 48–51.

ASAMI, R. & HAGA, R. 1983. A study of the population ecology of the Brown Dipper *Cinclus pallasii* at the Tokachi River System, Hokkaido. *Tori* 32, 75–94 (In Japanese with English summary).

ASH, J. S. 1960. A study of the Mallophaga of birds with particular reference to their ecology. *Ibis* 102, 93–110.

BAAKE, W. 1982. Ein Harzbach und seine Wasseramseln. *Falke* 29, 373–6.

BAGNALL-OAKLEY, R. P. 1954. Observations on a Black-bellied Dipper walking along the bottom of a stream. *Brit. Birds* 47, 28.

BAILLIE, S. R. & PEACH, W. 1992. Population limitation in Palaearctic–African migrant passerines. *Ibis* 134, Suppl. 1, 120–32.

BAKER, E. C. S. 1932. *The Nidification of Birds of the Indian Empire* Vol. 1. Taylor & Francis, London.

BAKER, E. C. S. 1942. *Cuckoo Problems* p. 186. H. F. & G. Witherby Ltd, London.

BAKUS, G. J. 1959a. Observations of the life history of the Dipper in Montana. *Auk* 76, 190–207.

BAKUS, G. J. 1959b. Territoriality, movements and population density of the Dipper in Montana. *Condor* 61, 410–52.

BALAT, F. 1960. A study on moulting in the Dipper *Cinclus cinclus* (L). *Zool. Listy* 9, 257–64 (In Czechoslovakian, with English summary).

BALAT, F. 1961. Colour variability in the Dipper, *Cinclus cinclus* (L), in Central Moravia and its taxonomic relations. *Zool. Listy.* 10, 135–46 (In Czechoslovakian, with English summary).

BALAT, F. 1962a. Studien über das Nisten und die Populationsdynamik der Wasseramsel (*Cinclus cinclus* L). *Sbornik Prednasek* S, 19–21. Prague.

BALAT, F. 1962b Distribution and movements of the dipper *Cinclus c. aquaticus* Bechst. on a creek and their changes during a year. *Zool. Listy* 11, 131–44 (In Czechoslovakian, with English summary).

BALAT, F. 1964. Breeding biology and population dynamics of the Dipper. *Zool. Listy*. 13, 305–20.

BANNERMAN, D. A. 1954. *The Birds of the British Isles* Vol. 3, pp. 396–7. Oliver & Boyd, Edinburgh & London.

BENNETT, G. F., WHITEWAY, M. & WOODWORTH-LYNES, C. 1982. *A Host-Parasite Catalogue of the Avian Haematozoa*. Memorial Unit of Newfoundland.

BENT, A. C. 1940. Life histories of North American cuckoos, goatsuckers, hummingbirds and their allies. *US Natl. Mus. Bull.* 176, 111–30.

BENT, A. C. 1948. Life histories of North american nuthatches, wrens, thrashers and their allies. *US Natl. Mus. Bull.* 195, 96–113.

BERGMAN, N. C. 1947. Ueber die Verhältnisse der Warmeokonomie der Thiere zu ihrer Grösse. *Gottinger Studien* 3, 595–708.

BIBBY, C. 1982. Polygyny and the breeding ecology of Cetti's Warbler *Cettia cetti*. *Ibis* 124, 288–301.

BIBBY, C. & THOMAS, D. K. 1984. Sexual dimorphism in size, moult and measurements of Cetti's Warbler *Cettia cetti*. *Bird Study* 31, 28–34.

BORGSTROM, E. 1991. Stromstarens *Cinclus cinclus* utbredning och hacknungsbiologi i Varmland. *Ornis Svecica* 1, 2.

BOU (BRITISH ORNITHOLOGISTS' UNION) 1971. *The Status of Brids in Britain and Ireland*, ed. D. W. Snow. Blackwell Scientific, Oxford.

BREHM, A. 1907. In ed. J. O. Bovig-Petersen, *Dyrenes liv*, p. 11, Copenhagen.

BREITENMOSER-WURSTEN, C. 1988. Zur Brutbiologie der Wasseramsel (*Cinclus cinclus*) im Saaneland (Berner Oberland, Schweize Nordalpen). *Okol. Vogel* 10, 119–50.

BREITENMOSER-WURSTEN, C. & MARTI, C. 1987. Verbreitung und Siedlungsdichte von Wasseramsel *Cinclus cinclus* und Bergstelze *Motacilla cinerea* im Saaneland (Berner Oberland). *Orn. Beob.* 84, 151–72.

BROADLEY, B. J. 1985. Dipper with one leg. *Scott. Birds* 13, 268.

BROWNLOW, H. G. 1949. The underwater movements of the Dipper. *Brit. Birds* 42, 69–73.

BROWNLOW, H. G. 1953. Underwater movements of the Dipper. *Brit. Birds* 46, 73–4.

BRYANT, D. M., HAILS, C. J. & PRYS-JONES, R. 1985. Energy expenditure by free-living Dippers (*Cinclus cinclus*) in winter. *Condor* 87, 177–86.

BRYANT, D. M. & TATNER, P. 1988. Energetics of the annual cycle of Dippers *Cinclus cinclus*. *Ibis* 130, 17–38.

BUDRIS, R. R. 1981a. On the biology and territorial inter-relationships of some aquatic passerines in the mountains of Badakhshan. *Ornitologiya* 16, 160–2 (In Russian).

BUDRIS, R. R. 1981b. The ecology of *Cinclus cinclus* and *Cinclus pallasii* in the mountains of middle Asia. *Biull. Mosk. Obs. Ispyt. Prir. Otd. Biol.* 86, 41–4 (In Russian, with English summary).

CAMPBELL, B. & LACK, E. (eds) 1985. *A Dictionary of Birds*. BOU. T. & A. D. Poyser, Calton.

CARLANDER, K. D. 1969. *Handbook of Freshwater Fishery Biology* Vol. 1. Iowa St. Univ. Press, Ames.

CARPENTER, K. 1924. A study of rivers polluted by lead mining in the Aberystwyth district of Cardiganshire. *Ann. Appl. Biol.* 12, 1–13.

CASTELINO, M. A. 1985. Mirlo de agua en Jujuy. *Nuestras Aves Asoc. Ornit. del Plata* 3, 26.

CHAMBERS, D. 1980. Foraging behaviour of the Dipper *Cinclus cinclus*. Unpubl. BSc. thesis, Plymouth Polytechnic.

CHAWORTH-MUSTERS, J. L. 1939. Some notes on the birds of the High Atlas of Moroccco. *Ibis* 14, 269–81.

COLLAR, N. J., GONZAGA, L. P., KRABBE, N., MADRONO NIETO, A., NARANJO, L. G., PARKER, T. A. & WIEGE, D. C. 1992. *Threatened Birds of the Americas. The ICBP/IUCN Red Data Book*, 3rd edn, Part 2. p. 1150. Smithsonian Institution Press, Washington DC.

COLE, E. A. 1990. A study to investigate the effect of organic pollution on populations of Dippers *Cinclus cinclus* and Grey Wagtails *Motacilla cinerea* on the River Wye, Derbyshire. Unpubl. MSc thesis, University of Manchester.

CORDIER, A. H. 1927. Some observations on the water Ouzel. *Auk* 44, 169–78.

COUSINS, S. A. 1985. Non-breeding territoriality of Dippers (*Cinclus cinclus*) and the influence of environmental conditions on their behaviour. Unpubl. BSc thesis, University of Stirling.

COWARD, T. A. 1969. *Birds of the British Isles and their eggs*. Ed./revised by J. A. G. Barnes. Frederick Warne & Co Ltd, London.

COWARD, T. A. & OLDHAM, C. 1910. *The Vertebrate Fauna of Cheshire and Liverpool Bay* Vol. 1. Witherby, London.

COWPER, C. N. L. 1973. Breeding distribution of Grey Wagtails, Dippers and Common Sandpipers on the Midlothian Esk. *Scottish Birds* 7, 302–6.

CRAIG, R. J. 1987. Divergent prey selection in two species of waterthrushes (*Seiurus*). *Auk* 104, 180–7.

CRAMP, S. (ed.) 1988. *Handbook of the Birds of Europe, the Middle East and North Africa: Birds of the Western Palearctic* Vol. 5. Oxford University Press, Oxford.

CREUTZ, G. 1952. Einemsen (anting) bei *Cinclus*. *J. Orn.* 93, 174.

CREUTZ, G. 1966. *Die Wasseramsel*. A. Ziensen, Wittenburg Lutherstadt.

CRICK, H., DUDLEY, C. & GLUE, D. 1993. Nest success in 1991. *BTO News* 187, 6–8.

CUMMINS, K. W. & WUYCHECK, J. C. 1971. Calorie equivalents for investigations in ecological energetics. *Mitteilungen der Internationalen Vereinigung für Theoretische und Angewandte Limnologie* 18, 1–148.

CURRY-LINDAHL, K. 1963. In Niels Blaedel, *Nordens fugle i farver*, Bd 1. Kopenhagen.

CZAPULAK, A. 1988. Distribution and breeding ecology of Dippers in Klodzko Land. *Birds of Silesia* 6, 97–116 (In Polish, English summary).

DA PRATO, S. 1981. The effect of spates on the feeding behaviour of Dippers. *Bird Study* 28, 60–2.

DA PRATO, S. & LANGSLOW, D. 1976. Breeding Dippers and their food supply. *Edinburgh Ring. Group Ann. Rept.* 4, 33–6.

DAULNE, J.-M. 1990. Distribution du cincle plongeur (*Cinclus cinclus*) dans le bassin de l'Aigne (province du Luxembourg). *Aves* 27, 27–38.

DEMENTIEV, G. P. & GLADKOV, N. A. (eds) 1954. *Birds of the Soviet Union* Vol. 6. Israel Progr. Sci. Transl., Jerusalem.

DE VOGUE, COMPTE G. 1934. Reflexions sur la marche du Cincle au fond de l'eau a propos de l'article de Monsieur Madon. *Alauda* 6, 526–8.

DEWAR, J. M. 1938.The dipper walking underwater. *Brit. Birds* 32, 103–6.

DICK, G. & SACKL, P. 1985. Untersuchungen zur Verbreitung, Siedlungsdichte und Nestplatzwahl der Wasseramsel (*Cinclus cinclus*) in Flußsystem des Kamp (Niederösterreich). *Okol Vogel* 7, 197–208.

DICK, G. & SACKL, P. 1988. Elterlicher Aufwand am Beispiel einer Wasseramselbrut (*Cinclus cinclus*). *Egretta* 31, 70–7.

DIESSELHORST, G. 1938. Zur Brutbiologie der Wasseramsel. *Beitr. Fortpfl. biol. Vogel* 14, 224–5.

DOBINSON, H. M. & RICHARDS, A. J. 1964. The effects of the severe winter of 1962/63 on birds in Britain. *Brit. Birds* 57, 373–434.

DREW, F. M. 1881. Field notes on some birds of San Juan County, Colorado. *Auk* (old series) 6, 85–91.

EALEY, D. M. 1977. Aspects of the ecology and behaviour of a breeding population of Dippers (*Cinclus mexicanus*) in southern Alberta. Unpubl. MSc thesis, 198 pp. Univ. Alberta, Edmonton.

EDWARDS, R. W. & BROOKER, M. P. 1982. *The Ecology of the Wye*. Junk, The Hague.

EFTELAND, S. 1975. Tidleg Hekking hja Fossekall (Early breeding in the dipper). *Sterna* 14, 185–9.

EFTELAND, S. 1976. Fossekallens dukketid og Dukkehyppighet (Duration and frequency of dives). *Sterna* 15, 11–8.

EFTELAND, S. 1983a. Skjor og krake pa fossekall-jakt (Magpies and Crows hunting Dipper) *Var Fuglefauna* 6, 52–3.

EFTELAND, S. 1983b. Uvanleg reirplassering, opphaldslokalitet osv. hos fossekall. *Var Fuglefauna* 6, 252–4.

EFTELAND, S. 1983c. Meir om fossekallens dukkeverksemd. *Var Fuglefauna* 6, 266–8.

EFTELAND, S. & KYLLINGSTAD, K. 1984. Nesting success in a SW-Norwegian dipper *Cinclus cinclus* population. *Fauna norv. Ser. C. cinclus* 7, 7–11.

EGGBRECHT, E. 1937. Brutbiologie der Wasseramsel (*Cinclus c. aquaticus* Bechst.). *J. F. Orn.* 636–76.

EHINGER, C. E. 1930. Some studies of the American Dipper or Water Ouzel. *Auk* 47, 487–98.

ELLIOTT, L. L. & PECK, S. 1980. Dipper swallowed by trout. *Wilson Bull.* 92, 524.

ETCHECOPAR, R. D. & HUE, F. 1964. Cincle plongeur (*Cinclus cinclus*) pp. 407–8, in *Les Oiseaux de l'Afrique*, ed. N. Bonbee. Cie, Paris.

EVENDEN, F. G. 1943. Food-washing habit of the Dipper. *The Condor* 45, 120.

EVERS, J. 1982. Occurrence of the Dipper *Cinclus cinclus* in Belgian and Dutch Limburg. *Wielewaal* 48, 237–9 (English summary).

FAIN, A., GREENWOOD, M. T. & MACFARLANE, D. 1991. Mites (Acari) found in the nests of the Dipper *Cinclus cinclus aquaticus* Bechstein, in Wales (British Isles). *Acarologia* 32, 193–204.

FJELDSA, J. & KRABBE, N. 1990. *Birds of the High Andes*. Zool. Museum, University of Copenhagen, Denmark. Apollo Books, Denmark.

FLINT, P. R. & STEWART, P. F. 1983. *The Birds of Cyprus*. Brit. Orn. Union. London.

FORREST, H. E. 1907. The Vertebrate Fauna of North Wales. Witherby, London.

FOWLER, J. A. & HODSON, D. P. 1991. Featherlice (Mallophaga) of the Dipper *Cinclus cinclus* in Central Wales. *Ring. & Migr.* 12, 43–5.

FRAGA, R. & NAROSKY, S. 1985. *Nidifacion de las aves Argentinas (Formicariidae a Cinclida)*. Associacion Ornitologica del Plata, Buenos Aires.

FREDRIKSSON, S. & GOTMARK, F. 1982. Ankomst, uppehall och rerelser av stromstarar *Cinclus c. cinclus* i Molndalsan under vintern 1976/77. *Var Fagelvarld* 41, 13–22 (In Swedish, with English summary).

FREETHY, R. 1980. Division of labour between Dippers building nests. *Brit. Birds* 73, 352.

FUCHS, E. 1970. Zur Brutbiologie der Wasseramsel (*Cinclus cinclus*). *Orn. Beob.* 67, 3–14.

GALBRAITH, H. 1979. Polygamy in the Dipper. *Scott. Birds* 10, 180–1.

GALBRAITH, H. 1988. Cited in Cramp (1988).

GALBRAITH, H. & BROADLEY, B. 1980. Biometrics and sexing of the British race of the Dipper. *Ring. & Migr.* 3, 62–4.

GALBRAITH, H., MITCHELL, A. B. & SHAW, G. 1981. The moult of the dipper in central Scotland. *Bird Study* 28, 53–9.

GALBRAITH, H. & TYLER, S. J. 1982. Movements and mortality of the Dipper as shown by ringing recoveries. *Ring. & Migr.* 4, 9–14.

GIBBONS, D. W., REID, J. B. & CHAPMAN, R. A. 1993. *The New Atlas of Breeding Birds in Britain and Ireland.* Published for the BTO, SOC and IWC by T. & A. D. Poyser, London.

GILINSKY, E. 1984. The role of fish predation and spatial heterogeity in determining benthic community structure. *Ecology* 65, 455–68.

GLADSTONE, H. S. 1910. *The Birds of Dumfriesshire.* Witherby, London.

GLUTZ VON BLOTZHEIM, U. N. 1962. *Die Brutvogel der Schweiz Wasseramsel*, pp. 409–12. Argauer Tagblatt (Arau).

GOLD, C. S. & DAHLSTEN, D. L. 1983. Effects of parasitic flies (*Protocalliphora* spp.) on nestlings of Mountain and Chestnut-backed chickadees. *Wilson Bull.* 95, 560–72.

GOODGE, W. R. 1959. Locomotion and other behaviour of the Dipper. *Condor* 61, 4–17.

GOODGE, W. R. 1960. Adaptations for amphibious vision in the Dipper (*Cinclus mexicanus*). *J. Morphol.* 107, 79–91.

GRAF, J. 1968. In *Animal Life of Europe*, ed. P. Michael. London and New York.

GREEN, J. C. 1970. Additional Dipper sightings in 1970. *Loon* 42, 137.

GREENHALGH, M. 1969. Dipper nesting in pot-hole. *Naturalist* 910, 84.

GREENWAY, J. C. & VAURIE, C. 1958. Remarks on some forms of *Cinclus* (Aves). *Breviora, Mus. Comp. Zool.* 89, 1–10.

GRINNELL, J. & STORER, T. I. 1924. *Animal Life in the Yosemite.* Univ. Calif. Press, California.

GRIST, W. R. 1934. Kingfisher robbing Dipper of food. *Brit. Birds* 27, 304–5.

GRUNWALD, K. 1973. Drei Bruten der Wasseramsel in einem Kasten. *Orn. Mitt.* 25, 78–9.

HAARTMAN, L. Von 1969. Nest site and evolution of polygyny in European passerine birds. *Ornis. Fenn.* 46, 1–12.

HAENSEL, J. 1977. Zum Vorkommen der Wasseramsel (*Cinclus cinclus aquaticus* Bechst.) im Harz. *Beitr. Vogelk.* 23, 9–30.

HAFTHORN, S. 1971. *Norges fugler.* Universitetsforlaget, Oslo–Bergen–Tromso.

HALSTEAD, J. A. 1988. American Dippers parasitised by blowfly larvae and the northern fowl mite. *Wilson Bull.* 100(3), 507–8.

HANEDA, K. & KOSHIKARA, S. 1969. Life History of the Pallas' Dipper (*Cinclus pallasii hondoensis*). *Bull. Inst. Nat. Educ.* Shiga Heights 8, 49–57 (Japanese with English summary).

HANN, H. W. 1950. Nesting behaviour of the American Dipper in Colorado. *Condor* 52, 49–62.

HARDEY, J., RAE, R. & RAE, S. 1978. Breeding success of Dippers in the Grampian region. *Grampian Ring. Group Rep.* 1, 23–5.

HEIM DE BALSAC, H. & MAYOUD, N. 1962. *Les Oiseaux de nord-ouest de l'Afrique.* Paul Lechavalier, Paris.

HESS, C. 1912. Vergleichende Physiologie des Gesichtssines. In H. Winterstein, *Handuch der vergleichenden Physiologie* Vol. 4. Gustav Fischer, Jena.

HEWSON, R. 1967. Territory, behaviour and breeding of the Dipper in Banffshire. *Brit. Birds* 60, 244–52.

HEWSON, R. 1969. Roosts and roosting habits of the Dipper. *Bird Study* 16, 89–100.

HOGSTEDT, G. 1980. Evolution of clutch size in birds: adaptive variation to territory quality. *Science* 210, 1148–50.

HOGSTEDT, G. 1981. Effects of additional food on reproductive success in the Magpie (*Pica pica*). *J. Anim. Ecol.* 50, 219–29.

HOLMBRING, J.-A. & KJEDEMAR, H. 1968. The dipper *Cinclus cinclus* in the province of Ostergotland, Sweden. *Var Fagelvarld* 27, 97–121 (In Swedish, with English summary).

HORN, W. 1985. Territory size of Dippers *Cinclus cinclus gularis* in relation to habitat type. Unpubl. BSc thesis, University of Edinburgh.

HORVATH, R. 1987. Dippers in Hungary. Paper given at Dipper Symposium at Rosenburg, Austria 5–6 September 1987.

HORVATH, R. 1988. Angaben über die Wasseramsel (*Cinclus cinclus*) in Ungarn. *Egretta* 31, 12–17.

HORVATH, R. & ANDRIKOVICS, S. 1991. A Vizirago (*Cinclus cinclus* L 1758) teli taplalekosszetetelerol (Winter foods of the dipper). In Hungarian with English summary. *Aquila* 98, 147–62.

HOSONO, R. 1978. Census of *Cinclus pallasii. Tori* 27, 21–6 (In Japanese, with English summary).

INGRAM, G. C. S. 1938. The dipper walking under water. *Brit. Birds* 32, 160.

INGRAM, G. C. S., SALMON, H. M. & TUCKER, B. W. 1938. The movements of the Dipper underwater. *Brit. Birds* 32, 58–63.

JACOB, J. P. 1988. Cincle plongeur, *Cinclus cinclus*, pp. 230–1, in *Atlas des oiseaux nicheurs de Belgique*, eds Devillers *et al.* Institut Royal des Sciences Naturelles de Belgiques, Brussels.

JAMES, F. C. 1970. Geographic size variation in birds and its relationship to climate. *Ecology* 51, 365–90.

JOHNSON, N. K. 1952. Dipper eaten by Brook Trout. *Condor* 55, 158.

JOHNSTON, V. R. 1943. An ecological study of nesting birds in the vicinity of Boulder, Colorado. *Condor* 45, 61–8.

JONES, J. W. & KING, G. M. 1952. The underwater activities of the Dipper. *Brit. Birds* 45, 400–1.

JOST, O. 1969. Uber die Bedeutung der Wasserscheiden beim Ortswechsel der Wasseramsel (*Cinclus cinclus aquaticus*). *J. Orn.* 110, 71–8.

JOST, O. 1970. Erfolgreiche Schutzmaßnahmen in dem Brutverein der Wasseramsel (*Cinclus cinclus*). *Angew Orn.* 3, 101–8.

JOST, O. 1972. Erwerb und Behandlung der Kocherfliegenlarven durch die Wasseramsel — *Cinclus cinclus. Luscinia* 41, 298–301.

JOST, O. 1975a. *Zur Okologie der Wasseramsel* (Cinclus cinclus) *mit besonderer Berücksichtigung ihrer Ernährung.* Bonner zool. Monograph. 6. Zoologisches Forschungsinstitut und Museum Alexander Koenig, Bonn.

JOST, O. 1975b. Fisch-otolithen in Speiballen der Wasseramsel. *Natur Mus.* 105, 283–6.

JOURDAIN, F. C. R. 1911. Early breeding of the Kingfisher and Dipper in Denbighshire. *Brit. Birds* 4, 318–9.

KAISER, A. 1985. Zur Verbreitung und Bestandssitiation der Wasseramsel (*Cinclus c. aquaticus*) in Rheinhessen, Rheingau und ostlichem Hunsbruck. *Okol. Vogel* 7, 185–96 (Distribution and status of Dippers).

KAISER, A. 1988. Zur Populationsdynamik der Wasseramsel (*Cinclus cinclus*) in Rheinhessen und angrenzenden Gebieten. *Egretta* 31, 18–37.

KEAR, J. & BURTON, P. J. K. 1971. The food and feeding apparatus of the Blue Duck *Hymenolaimus malacorhynchus. Ibis* 113, 483–93.

KEICHER, K. 1983. Untersuchungen zur Aktivitätsrhythmik und zum Schlafplatzverhalten der Wasseramsel (*Cinclus aquaticus*) am oberen Neckar (Wurttemberg). *Okol Vogel* 5, 203–16.

KEITH, S., URBAN, E. K. & FRY, C. H. 1992. *The Birds of Africa* Vol. 4. Academic Press, London.

KEMPF, C. 1977. *Oiseaux de Alsace.* Librairie Istra, Strasbourg.

KENDEIGH, S. C., DOL'NIK, V. R. & GAVRILOV, V. M. 1977. Avian Energetics, pp. 127–204, in *Granivorous Birds in Ecosystems*, ed. J. Pinowski & S. C. Kendeigh. Cambridge University Press, Cambridge.

KING, W. B. 1978–79. *Red Data Book, 2 Aves* 2nd edn. International Union for Conservation of Nature and Natural Resources, Morges, Switzerland.

KING, W. B. 1981. *Endangered Birds of the World. The ICBP Red Data Book.* Smithsonian Inst. Press, Washington DC.

KLEIN, W. & SCHAAK, K. 1972. Zur Siedlungsokologie der Wasseramsel (*Cinclus cinclus*) im Wassereinzugsgebiet der Kinzig/Hessen. *Luscinia* 41, 277–97.

KOSKIMIES, P. 1989. Distribution and Numbers of Finnish breeding birds. Appendix to Suomen lintuatlas SLY:n Lintutieto Oy. Helsinki.

KOVSHAR, A. F. 1979. *Pevchie ptitsy v subvysokogor'e Tyan Shanya* (Song-birds in the subalpine of Tien-Shan), pp. 43–53. Alma-Ata, USSR.

LACHENMAYER, E., KUNZE, P. & HOLZINGER, J. 1985. Schwermetalle in Nahrungstieren und Eiern der Wasseramsel (*Cinclus c. aquaticus*) und Gebirgsstelze (*Motacilla cinerea*) im Kirchheimer Raum. *Okol. Vogel* 7, 327–52 (Heavy metals in food and eggs of Dippers and Grey Wagtails in SW Germany).

LACK, D. 1947. The significance of clutch size I and II. *Ibis* 89, 302–52.

LACK, D. 1948a. Natural selection of family size in the starling. *Evolution* 2, 95–110.

LACK, D. 1948b. The significance of clutch size III. *Ibis* 90, 25–45.

LACK, D. 1954. *The Natural Regulation of Animal Numbers.* Clarendon Press, Oxford.

LAGERSTROM, M. 1979. Wintering Dippers (*Cinclus cinclus*) at Pirkanmaa, southern Finland, in the winters of 1963/64–1977/78. Paper presented 3rd meeting work group Dipper res. Scand. Oslo 1979.

LASSO, S. V. 1988. Historia Natural Y Distribucion del Mirlo aquatica Gorribeanco (*Cinclus leucocephalus leuconotus*) en la hoya del Rio Guayllabamba, Ecuador.

LE HARDY DE BEAULIEU, F. 1973. Nids accole due Cincle plongeur (*Cinclus cinclus*) et du Troglodyte (*Troglodytes troglodytes*). *Aves* 10, 208.

LEHIKOINEN, E. & HAKALA, J. 1988. Variation in weight of migratory Dippers *Cinclus cinclus* in their Finnish winter quarters. *Bird Study* 35, 101–8.

LEWIS, G. & WILLIAMS, G. 1984. *Rivers and Wildlife Handbook.* RSPB and RSNC.

LEWIS, J. M. S. 1991. Polygamy in the Dipper. *The Welsh Bird Report* 5, 87, Welsh Ornithological Society.

LIPPENS, L. & WILLE, H. 1972. *Atlas des Oiseaux de Belgique et d'Europe occidentale.* Lannoo, Tielt.

LUNDBERG, P., BERGMAN, A. & OLSSON, H. 1981. On the ecology of wintering Dippers (*Cinclus cinclus*) in northern Sweden. *J. Orn.* 122, 163–72.

MACEY, R. W. & STRONG, G. L. 1967. *Laterotrema cascadensis* sp. N. (Trematoda: Stomylotrematidae) from the dipper *Cinclus mexicanus. J. Parasit.* 53, 584–6.

MADON, M. P. 1934. Notes sur deux especes particulierement interessantes. Le cincle et l'engoulevent. *Alauda* 6, 47–65.

MALLON, D. P. 1987. The winter birds of Ladakh. *Forktail* 3, 27–41.

MANUEL, F. 1949. Une Buse egorge un Cincle pris au filet. *Nos Oiseaux* 20, 149.

MARCHANT, J. H., HUDSON, R., CARTER, S. P. & WHITTINGTON, P. 1990. *Population Trends in British Breeding Birds.* BTO, Tring.

MARCHANT, J. H. & HYDE, P. M. 1979. Population changes for waterways birds, 1974–1978. *Bird Study* 26, 227–39.

MARCHANT, J. H. & HYDE, P. M. 1980. Population changes for waterways birds for 1978–79. *Bird Study* 27, 179–82.

MARTENS, J. & GEDULDIG, G. 1990. Acoustic adaptations of birds living close to Himalayan torrents. Proc. Int. 100 DO-G Meeting, *Current Topics Avian Biol.*, Bonn 1998, pp. 123–31.

MASON, C. F. 1991. *The Biology of Freshwater Pollution.* Longmans, London.

MASON, T. B. 1953. Unusual nesting-site of Dipper. *Brit. Birds* 46, 35–6.

MAWBY, P. J. 1961. A study of the Dipper (*Cinclus gularis*) in the Sedbergh area. *Naturalist* 87, 45–9.
MAWSON, G. P. 1979. Parasitic flies. A brief guide to Hippoboscid flat flies found on birds. *Sorby Breck Ringing Group Report* pp. 23–26.
MEAD, C. J. & CLARK, J. A. 1988. Report on bird ringing in Britain and Ireland for 1987. *Ring. & Migr.* 9, 169–204 (p. 198).
MEAD, C. J. & CLARK, J. A. 1989. Report on bird ringing for 1988. *Ring. & Migr.* 10, 158–96 (p. 188).
MEAD, C. J. & CLARK, J. A. 1991. Report on bird ringing for Britain and Ireland for 1990. *Ring. & Migr.* 12, 139–76 (p. 170).
MEAD, C. J. & CLARK, J. A. 1993. Report on bird ringing in Britain and Ireland for 1991. *Ring. & Migr.* 14, 1–72 (p. 59).
MEIKLEJOHN, M. F. M. 1948. Birds of Teheran and Alborz. *Ibis* 90, 83.
MERIKALLIO, E. 1958. Finnish birds: their distribution and numbers. *Fauna Fenn.* 5, 118–19.
MITCHELL, P. A. 1968. The food of the dipper (*Cinclus mexicanus* Swainson) on two western Montana streams. Unpubl. MA thesis. Univ. of Montana, Missoula.
MOLL, G. 1977. Zahlung von Vögeln an der Ruhr (Eifel) und ihren Nebenbachen. *Charadrius* 13, 15–6.
MOLLER, A. P. 1989. Population dynamics of a declining swallow *Hirundo rustica* population. *J. Anim. Ecol.* 58, 1051–63.
MONIG, R. 1985. Zur Indikatorqualität der Wasseramsel (*Cinclus c. aquaticus*) — Ruckstandssanalytischer Nachweis von polychlorierten Biphenylen (PCBs) in Eiern Fliessgewasser bewohnender Vogel. *Okol. Vogel* 7, 353–8 (The use of Dippers as a bio-indicator of residues of PCBs in the eggs of river birds).
MOODY, C. 1955. Display flight of Dipper. *Brit. Birds* 48, 184.
MOON, H. J. 1923. Dippers nesting away from water. *Brit. Birds* 17, 59.
MORIN, P. J. 1984. The impact of fish exclusion on the abundance and species composition of larval odontates: results of short-term experiments in a North Carolina farm pond. *Ecology* 65, 53–60.
MORK, K. 1975. Bigami-og to kull i same sesong-pavist hos fossekall. *Sterna* 14, 131–4.
MORRIS, F. O. 1896. *A History of British Birds* Vol. III. Witherby, London.
MOSS, D. 1975. Movements and sexing of roosting Dippers. *Edinburgh Ringing Group Annual Report* 3, 8–15.
MUELHAUSEN, M. H. 1970. First Dipper record for Minnesota. *Loon* 42, 136.
MUIR, J. 1894. *The Mountains of California*. The Century Co., New York.
MURPHY, P. 1972. An analysis of the variation in winter territory length of the Dipper. Unpubl. BSc thesis, University of Edinburgh.
MURRAY, J. B. 1967. Dippers diving in icebound loch. *Scott. Birds* 4, 450–1.
MURRISH, D. E. 1970a. Responses to diving in the Dipper *Cinclus mexicanus*. *Comp. Biochem. Physiol.* 34, 853–8.
MURRISH, D. E. 1970b. Responses to temperature in the Dipper *Cinclus mexicanus*. *Comp. Biochem. Physiol.* 34, 859–69.
NEWTON, I. 1991. Long-term monitoring of organochlorine and mercury residues in some predatory birds in Britain. *Acta XX Congressus Ornithologici*, IV, 2487–93.
NEWTON, S. F. 1993. Body condition of a small passerine bird: ultrasonic assessment and significance in overwinter survival. *J. Zool. Lond.* 229, 561–80.
NICOLL, M. 1971. Dippers and Grey Wagtails on the Elliot Water *Tay Ring. Group Report* 5, 5.
NYHLEN, G. 1952. Stromstare (*Cinclus cinclus*) och hussvala (*Delichon urbica*) i samma bo. *Var Fagelvarld* 11, 33–4.

OELKE, H. 1975. Wasseramsel (*Cinclus cinclus*) und Gebirgsstelze (*Motacilla cinerea*) im Westharz (Bestandsaufnahme 1973). *Vogelk. Ber. Nidersachs* 7, 19–31.

O'HALLORAN, J. O., GRIBBIN, S. D., TYLER, S. J. & ORMEROD, S. J. 1990. The ecology of dippers *Cinclus cinclus* (L) in relation to stream acidity in upland Wales: time-activity budgets and energy expenditure. *Oecologia* 85, 271–80.

O'HALLORAN, J. O., SMIDDY, P. & O'MAHONY, B. 1993. Biometrics, growth and sex ratios amongst Irish Dippers *Cinclus cinclus hibernicus*. *Ring. & Migr.* 13, 152–61.

ORENSTEIN, R. J. 1975. Observations and comments on two stream-adapted birds of Papua New Guinea. *Bull. BOC* 95, 161–5.

ORIANS, G. 1969. On the evolution of mating systems in birds and mammals. *Am. Nat.* 103, 589–603.

ORMEROD, S. J. 1985a. The diet of breeding Dippers *Cinclus cinclus* and their nestlings in the catchment of the River Wye, mid-Wales: a preliminary study by faecal analysis. *Ibis* 127, 316–31.

ORMEROD, S. J. 1985b. Optimal foraging by breeding Dippers on a Lancashire hill stream. *Naturalist* 110, 99–103.

ORMEROD, S. J. 1988. The micro- and macrodistribution of macroinvertebrates in the catchment of the River Wye: the result of biotic or abiotic factors? *Freshwater Biology* 20, 241–7.

ORMEROD, S. J., ALLINSON, N., HUDSON, D. & TYLER, S. J. 1986. The distribution of breeding Dippers (*Cinclus cinclus* (L) Aves) in relation to stream acidity in upland Wales. *Freshwat. Biol.* 16, 501–7.

ORMEROD, S. J., BOILSTONE, M. A. & TYLER, S. J. 1985. Factors influencing the abundance of breeding Dippers *Cinclus cinclus* in the catchment of the River Wye, mid-Wales. *Ibis* 127, 332–40.

ORMEROD, S. J., BULL, K. R., CUMMINS, C. P., TYLER, S. J. & VICKERY, J. A. 1988a. Egg mass and shell thickness in Dippers *Cinclus cinclus* in relation to stream acidity in Wales and Scotland. *Environ. Pollut.* 55, 107–21.

ORMEROD, S. J., EFTELAND, S. V. & GABRIELSEN, L. E. 1987. The diet of breeding dippers *C. cinclus cinclus* and their nestlings in south western Norway. *Holarctic Ecology* 10, 201–5.

ORMEROD, S. J., O'HALLORAN, J., GRIBBIN, D. D. & TYLER, S. J. 1991. The ecology of dippers (*Cinclus cinclus* (L)) in relation to stream acidity in upland Wales: breeding performance, calcium physiology and nestling growth. *J. Appl. Ecol.* 28, 419–33.

ORMEROD, S. J. & PERRY, K. W. 1985. The diet of breeding Dippers and their nestlings in north-west Ireland. *Irish Birds* 3, 90–5.

ORMEROD, S. J. & TYLER, S. J. 1986. The diet of Dippers *Cinclus cinclus* wintering in the catchment of the River Wye, Wales. *Bird Study* 33, 36–45.

ORMEROD, S. J. & TYLER, S. J. 1987a. Aspects of the breeding ecology of Welsh Grey Wagtails *Motacilla cinerea*. *Bird Study* 34, 43–51.

ORMEROD, S. J. & TYLER, S. J. 1987b. Dippers (*Cinclus cinclus*) and Grey Wagtails (*Motacilla cinerea*) as indicators of stream acidity in Upland Wales. *ICBP Technical Publ.* 6, 191–208.

ORMEROD, S. J. & TYLER, S. J. 1990a. Population characteristics of Dipper *Cinclus cinclus* roosts in mid and south Wales. *Bird Study* 37, 165–70.

ORMEROD, S. J. & TYLER, S. J. 1990b. Environmental pollutants in the eggs of Welsh Dippers *Cinclus cinclus*: a potential monitor of organochlorine and mercury contamination in upland rivers. *Bird Study* 37, 171–6.

ORMEROD, S. J. & TYLER, S. J. 1991a. Exploitation of prey by a river bird, the dipper *Cinclus cinclus* (L), along acidic and circumneutral streams in upland Wales. *Freshwater Biology* 25, 105–16.

ORMEROD, S. J. & TYLER, S. J. 1991b. The influence of stream acidification and riparian land use on the feeding ecology of Grey Wagtails *Motacilla cinerea* in Wales. *Ibis* 133, 53–61.

ORMEROD, S. J. & TYLER, S. J. 1992. Patterns of contamination by organochlorines and mercury in the eggs of two river passerines in Britain and Ireland with reference to individual PCB congeners. *Environ. Pollut.* 76, 223–43.

ORMEROD, S. J. & TYLER, S. J. 1993a. Further studies on local and regional patterns in the organochlorine content of eggs of a river bird, the dipper *Cinclus cinclus*. *Bird Study* 40, 97–106.

ORMEROD, S. J. & TYLER, S. J. 1993b. Birds as indicators of changes in water quality, pp. 179–216. In *Birds as Monitors of Environmental Change*, eds J. J. D. Greenwood & R. J. Furness. Chapman Hall, London.

ORMEROD, S. J. & TYLER, S. J. 1993c. The adaptive significance of brood size and time of breeding in the Dipper *Cinclus cinclus* (Aves: Passeriformes) as seen from post-fledging survival. *J. Zool.* 231, 371–81.

ORMEROD, S. J., TYLER, S. J. & LEWIS, J. M. S. 1985. Is the breeding distribution of Dippers influenced by stream acidity? *Bird Study* 32, 33–9.

ORMEROD, S. J., TYLER, S. J. & LEWIS, J. M. S. 1986. Biometrics, growth and sex ratios amongst Welsh Dippers *Cinclus cinclus*. *Ring. Migr.* 7, 61–70.

ORMEROD, S. J., TYLER, S. J., PESTER, S. J. & CROSS, A. V. 1988b. Censussing distribution and population of birds along upland rivers using measured ringing effort: a preliminary study. *Ring. & Migr.* 9, 71–82.

PENOT, J. 1948. Notes biologique sur le Cincle plongeur. *L'Oiseau* 18, 141–51.

PERIS, S. J., GONZALEZ-SANCHEZ, N.,CARNERO, J. I., VELASCO, J. C. & MASA, A. I. 1991. Algunas factores que inciden en la densidad y poblacion del Mirlo acuatica (*Cinclus cinclus*) en el centro-occidente de la Peninsula Iberica. *Ardeola* 38, 11–20.

PERRINS, C. & BIRKHEAD, T. 1983. *Avian Ecology*. Glasgow, Blackie.

PERRY, K. W. 1983. Population changes of the Dipper in north-west Ireland. *Irish Birds* 2, 272–7.

PERRY, K. W. 1986. *The Irish Dipper*. Published privately.

PERRY, R. 1976. Dippers on the Spey. *Wildlife* 18, 10–4.

PREUSS, N. O. 1959. Vandstaerens (*Cinclus cinclus* (L)) udbrelse og forekomst in Danmark, med bemaerkninger om dens opforsel i vinterkvarteret. *Dansk. Orn. Foren Tidsskr.* 53, 1–19.

PRICE, F. E. & BOCK, C. E. 1973. Polygyny in the Dipper. *Condor* 75, 457–9.

PRICE, F. E. & BOCK, C. E. 1983. Population ecology of the Dipper (*Cinclus mexicanus*) in the Front Range of Colorado. *Studies in Avian Biology* No.7. Cooper Ornithological Society.

RAE, R. & HARDEY, J. 1979. Nestboxes for Dippers. *Grampian Ring. Group Report* 2, 24–5.

RANKIN, M. N. & RANKIN, D. H. 1940. The breeding of the Irish Dipper. *Irish Nat. J.* 7, 273–2.

RHEINWALD, G. 1982. *Brutvogelatlas der Bundesrepublik Deutschland* Bonn.

RICHTER, H. 1953. Zur Lebensweise der Wasseramsel. Erster Teil: Der Ortswechsel. *J. Orn.*94, 68–72.

RICHTER, H. 1954a. Zur Variabilität der Farbung sachsischer Wasseramseln (*Cinclus cinclus aquaticus*). *J. Orn.* 95, 55–7.

RICHTER, H. 1954b. Zur Mauser der Wasseramsel, *Cinclus c.aquaticus* (Bechstein). *Beitr. Vogelk.* 3, 251–8.

ROBSON, J. E. 1969. Dipper flying into barbed wire fence. *Brit. Birds* 62, 498.

ROBSON, R. W. 1956. The breeding of the Dipper in north Westmoreland. *Bird Study* 3, 170–80.

ROCKENBAUCH, D. 1985a. Wasseramsel (*Cinclus c. aquaticus*) und Zivilisation — am Beispiel des Fluss-Systems der Fils (Schwabische Alb). *Okol. Volgel* 7, 171–84.

ROCKENBAUCH, D. 1985b. Geschlechts- und Altersbestimmung bei der Wasseramsel (*Cinclus c. aquaticus*). *Okol. Vogel* 7, 363–77.

ROOS, P. 1982. Voorkomen van de Waterspreeuw (*Cinclus cinclus*) in Nederland. *Limosa* 55, 3–8.

ROSKELL, J. 1982. Anting behaviour by Dipper. *Devon Birds* 35, 90–1.

ROTHSCHILD, M. & CLAY, T. 1952. *Fleas, Flukes and Cuckoos. A study of Bird Parasites*. Collins, London.

ROUND, P. D. & MOSS, M. 1984. The waterbird population of three Welsh rivers. *Bird Study* 31, 61–8.

SACHER, G. 1980. Bigamie bei der Wasseramsel *Cinclus cinclus* (L). *Thur. Orn. Mitt.* 26, 17–20.

SACKL, P. & DICK, G. 1988. Zur Brutbiologie der Wasseramsel (*Cinclus cinclus*) im Flußsystem des Kamp, Niederösterreich. *Egretta* 31, 56–69.

SALVADOR, S., NAROSKY, S. & FRAGA, R. 1986. First description of the nest and eggs of the Red-throated Dipper in northwestern Argentina. *Gerfaut* 76, 63–6.

SANTAMARINA, J. 1988. Distribucion de los mamiferos y aves de la cuenca media del Rio Ulla (Galicia) en relacion con los recursos y la calidad de las aguas. Conference proc. Coloquio Luso-Espanol sobre ecologia das bacias hydrograficas e recursos zoologicos. Porto, Portugal, 23–25 March 1988.

SANTAMARINA, J. 1990. Alimentacion del mirlo acuatica (*Cinclus cinclus*) en rios de Galicia. *Misc. Zool.* 14, 207–15.

SCHIFFERLI, L. GEROUDET, P. & WINKLER, R. 1980. *Verbreitungsatlas der Brutvögel der Schweiz*. Schweizerische Vogelwarte, Sempach.

SCHMID, W. 1985a. Daten zur Brutbiologie der Wasseramsel (*Cinclus c. aquaticus*) im Bachsystem der Lauter und Lindach im Landkreis Esslingen, Nordwurttemberg. *Okol. Vogel* 7, 225–38.

SCHMID, W. 1985b. Analyse der Ortsbewegung von nestjung beringten Wasseramseln (*Cinclus c. aquaticus*) im ersten Lebensjahr. *Okol. Vogel* 7, 359–62.

SCHMID, W. 1985c. Starker Befall durch die Nordliche Vogelmilbe (*Ornithonyssus sylviarum* Canestrini et Fanzago 1877) an Nestlingen der Wasseramsel (*Cinclus c. aquaticus*). *Okol. Vogel.* 7, 423–6.

SCHMID, W. & SPITZNAGEL, A. 1985. Der sexuelle Grossendimorphismus süddeutscher Wasseramseln (*Cinclus c. aquaticus*): Biometrie, Funktion und mögliche Ursachen. *Okol. Vogel* 7, 379–408.

SCHNEIDER, F. 1985. Erste Daten zur Winterverbreitung der Wasseramsel. *Vogelwelt* 74, 128–33.

SCHONWETTER, M. 1971. *Handbuch der Okologie*. Lief. Vogelzug Verlag, Berlin.

SCHUSTER, L. 1953. Einige Bemerkungen zur Biologie der Wasseramsel. *Vogelwelt* 74, 128–33.

SERLE, W. & BRYSON, D. 1935. Distribution and numbers of the Dipper on the North and South Esks (Midlothian). *Brit. Birds* 28, 327–31.

SHARROCK, J. T. R. 1976. *The Atlas of Breeding Birds in Britain and Ireland* BTO, IWC. T. & A. D. Poyser, Calton.

SHAW, G. 1978 The breeding biology of the Dipper. *Bird Study* 25, 149–60.

SHAW, G. 1979. Functions of Dipper roosts. *Bird Study* 26, 171–8.

SHIMELD, M. 1984. Unusual dipper nest site. *Scottish Birds* 13, 117.

SHOOTER, P. 1969a. Undomed nests of Dippers. *Brit. Birds* 62, 240.

SHOOTER, P. 1969b. Two Dippers' nests only 18 inches apart. *Brit. Birds* 62, 446–7.

SHOOTER, P. 1970. The Dipper population of Derbyshire, 1958–68. *Brit. Birds* 63, 158–63.

SIBLEY, C. G. 1970. *Bull. Peabody Nat. Hist. Mus.* Vol. 32.

SIBLEY, C. G. 1972. The relationship of certain families of passerine birds. *Proc. 15th Int. Orn. Conngr. Leiden.*

SIBLEY, C. G., CORBIN, C. W., AHLQUIST, J. E. & FERGUSON, A. 1974. In *Biochemical and Immunological Taxonomy of Animals*, pp. 80–176, ed. C. A. Wright. Academic Press, London.

SINTON, R. 1988. Proximate factors determining local-scale breeding territory size of the Dipper (*Cinclus c. gularis*) on the upper Wharfe, North Yorkshire. BSc project. Univ. of Leeds.

SMITH, R. P. & ORMEROD, S. J. 1986. The diet of moulting Dippers *Cinclus cinclus* in the catchment of the Welsh River Wye. *Bird Study* 33, 138–9.

SNOW, D. W. 1967. *A Guide to Moult in British Birds*. BTO Field Guide No. 11.

SOKOLOWSKI, J. 1964. Die wasseramsel, (*Cinclus cinclus* L) im Tatragebirge). *Przeglad Zoologiczny* 8, 349–59 (In Polish).

SPITZNAGEL, A. 1985a. Lausfliegen (Hippoboscidae, Diptera) und Milben (Acari) als ektoparasiten der Wasseramsel (*Cinclus c. aquaticus*) *Okol. Vogel.* 7, 421–2.

SPITZNAGEL, A. 1985b. Untersuchungen über den Befall der Wasseramsel (*Cinclus c. aquaticus*) mit Federlingen (Phthiraptera: Mallophaga). *Okol. Vogel.* 7, 409–20.

SPITZNAGEL, A. 1985c. Jahreszeitliche Veränderungen im Nahrungsangebot und in der Narungswahl der Wasseramsel (*Cinclus c. aquaticus*). *Okol. Vogel.* 7, 239–326 (Seasonal variation in food supply and food choice of the Dipper).

SPITZNAGEL, A. 1988. Strategien des Nahrungserwerbes bei der Wasseramsel (*Cinclus c. aquaticus*): eine Einführung. *Egretta* 31, 42–55.

STABLER, R. M., HOLT, P. A. & KITZMILLER, N. J. 1966. *Trypanosoma avium* in the blood and bone marrow from 677 Colorado birds. *J. Parasit.* 52, 1141–4.

STABLER, R. M. & KITZMILLER, N. J. 1970. Hematozoa from Colorado Birds. 111. Passeriformes. *J. Parasit.* 56, 12–6.

STAEDTLER, K. & BREMSHEY, K. 1988. Bestandsentwicklung der Wasseramsel (*Cinclus cinclus*) durch Nisthilfen im Raum Schwerte/Ruhr. *Egretta* 31, 38–41.

STANFORD, J. K. & MAYR, E. 1941. The Vernay-Cutting Expedition to Northern Burma. *Ibis* 5, 218–9.

STEIGER, J. A. 1940. Dipper, wilderness dweller. *Bird-lore* 42, 10–4.

STEINEGER, L. 1885. *Standard Nat. Hist. IV, Birds* p. 505.

STEINEGER, L. 1905. Birds of the genus *Cinclus* and their geographical distribution. *Smithsonian Misc. Coll.* 47, 421–30.

STEPANYAN, L. S. 1962. Observations on Brown Dipper in the central Tien Shan. *Ornitologiya* 5, 243–8.

STEPHENS, M. N. 1957. The Otter Report, pp. 56–57. Univ. Fed. An. Welfare.

SUDHAUS, W. 1972. Zur Verhaltensbiologie der Wasseramsel (*Cinclus cinclus*). *Orn. Mitt.* 24, 231–6.

SULLIVAN, J. O. 1965. 'Flightlessness' in the Dipper. *Condor* 67, 535–6.

SULLIVAN, J. O. 1966. A Dipper nest away from water. *Condor* 68, 107.

SULLIVAN, J. O. 1973. Ecology and behavior of the dipper: Adaptations of a passerine to an aquatic environment. PhD thesis, University of Montana, Missoula.

SUNQUIST, M. E. 1976. Territory size and nesting habits of Brown dippers *Cinclus pallasii*. *Ibis* 118, 577–8.

SVENSSON, L. 1992. *Identification Guide to European Passerines*. Stockholm, Sweden.

SWAINSON, C. 1886. *The Folklore and Provincial Names of British Birds*. Elliot Stock & Folklore Society, London.

TAYLOR, A. G. 1929. Dipper's nest on a roof. *Brit. Birds* 23, 163–4.

THORP, J. H. & BERGEY, E. A. 1981. Field experiments on responses of a freshwater benthic macroinvertebrate community to vertebrate predators. *Ecology* 62, 365–75.

THUT, R. N. 1970. Feeding habits of the Dipper in south western Washington. *Condor* 72, 234–5.

TROCHET, B. 1967. Un nid de cincle sur un arbre. *Jean le Blanc* 6, 100–1.

TYLER, G. 1990. The effect of farm pollution of rivers on birds. Unpubl. RSPB Report, Sandy, Beds.
TYLER, S. J. 1972. The breeding biology of the Grey Wagtail. *Bird Study* 19, 69–80.
TYLER, S. J. 1991.The impact of mink on waterbirds. Unpubl. paper RSPB, Newtown, Powys.
TYLER, S. J. 1994. The Yungas of Argentina: in search of Rufous-throated Dippers *Cinclus schulzi. Cotinga* 2, 38–41.
TYLER, S. J. & ORMEROD, S. J. 1985. Aspects of the breeding biology of Dippers in the southern catchment of the River Wye, Wales. *Bird Study* 33, 164–9.
TYLER, S. J. & ORMEROD, S. J. 1986. Dipper, pp. 306–7. In *The Atlas of Wintering Birds in Britain and Ireland*, ed. P. Lack. BTO/IWC. T. & A. D. Poyser, Calton.
TYLER, S. J. & ORMEROD, S. J. 1991. Aspects of the biology of Dippers *Cinclus cinclus minor* in the Atlas Mountains of Morocco outside the breeding season. *Bonn Zool. Beitr.* 42, 35–45.
TYLER, S. J. & ORMEROD, S. J. 1993. The Dipper. In *The New Atlas*, ed. R. Gregory. BTO. T. & A. D. Poyser, Calton.
TYLER, S. J. & ORMEROD, S. J. 1994. The ecology of river birds in Nepal: some implications of land use changes. *Forktail* 9, 59–82.
TYLER, S. J., ORMEROD, S. J. & LEWIS, J. M. S. 1990. Breeding and natal dispersal amongst Welsh Dippers. *Bird Study* 37, 18–23.
TYLER, S. J. & TYLER, L. In press. Some observations of Rufous-throated Dippers *Cinclus schulzi* in northwest Argentina. *Bird Conservation International.*
ULFSTRAND, S. & HOGSTEDT, G. 1976. Hur manga faglar hackar i Sverige? (How many birds breed in Sweden?) *Anser* 15, 1–32.
VADER, W. 1971. Dippers feeding on marine invertebrates. *Brit. Birds* 64, 456–8.
VERNER, J. & WILSON, M. F. 1966. The influence of habitats on mating systems of North American passerine birds. *Ecology* 47, 143–7.
VICKERY, J. 1988. The effects of surface water acidification on riparian birds — with particular reference to the Dipper. Unpubl. DPhil thesis, Univ. of Oxford.
VICKERY, J. 1991.Breeding density of Dippers *Cinclus cinclus*, Grey Wagtails *Motacilla cinerea* and Common Sandpipers *Actitis hypoleucos* in relation to the acidity of streams in south west Scotland. *Ibis* 133, 178–85.
VITTERY, A., PORTER, R. F. & SQUIRES, J. E. (eds). 1971. *Checklist of the Birds Of Turkey*. Orn. Soc. Turkey. Sandy, Beds.
VOGT, W. 1944. Uber die Territorien der Wasseramsel *Cinclus cinclus* (L) im Winter 1943/44 an der Aare bei Bern. *Orn. Beob.* 41, 36–43.
VOLLNHOFER, P. 1906. Uber die fischwirtschaftliche Bedeutung der Wasseramsel (*Cinclus cinclus* L). *Erdescz. Kiserl.* 8, 1–81.
VOOUS, K. H. 1960. *Atlas of European Birds*. Nelson, London.
WAGNER, S. 1985. Zur Situation der Wasseramsel (*C. c. aquaticus*) in Karnten (Südösterreich). *Okol. Vogel.* 7, 209–14.
WARD, R. C. 1968. Some runoff characteristics of some British rivers. *J. Hydrol. (Neth.)* 6, 358–72.
WATSON, D. 1972. *Birds of Moor and Mountains*. Scottish Academic Press. Edinburgh & London.
WIATR, B. 1969. Census of Dippers in the Tatra National Park. *Notatki orn.* 10, 31–4 (Polish, with English summary).
WILLIAMS, H. H. 1984. Dippers using mud as base for nest. *Brit. Birds* 77, 210–1.
WILSON, J. D. 1992. A probable case of sexually-selected infanticide in the dipper. *Ibis* 134, 188–90.
WINK, M. & GERSTBERGER, P. 1977. Der Bestand von Wasseramsel (*Cinclus cinclus*) und Eisvogel (*Alcedo atthis*) im Flußsystem der Ahr (Vergleich 1964–1976). *Charadrius* 13, 8–14.

WITHERBY, H. F., JOURDAIN, F. C. R., TICEHURST, N. F. & TUCKER, B. W. 1948. *The Handbook of British Birds*, Vol. 11. Witherby, London.
YALDEN, D. W. 1986. The diet, food availability and habitat selection of breeding Common Sandpipers *Actitis hypoleucos*. *Ibis* 128, 23–6.
YAMAGUTI, S. 1939. Studies on the helminth fauna of Japan, Part 25. Trematodes of birds. IV. *Jap. J. Zool.* 8, 129–210.
YEATMAN, L. J. 1976. *Atlas des oiseaux nicheurs de France*. Paris.
YOERG, S. I. & O'HALLORAN, J. 1991. Dipper nestlings fed by a Gray Wagtail. *Auk* 108, 427–9.
ZANG, H. 1981. Zur Brutbiologie und Hohenverbreitung der Wasseramsel (*Cinclus c. aquaticus*) im Harz. *J. Orn.* 122, 153–62.
ZANG, H. 1985. Hangexposition und Brutbiologie von Wasseramsel (*Cinclus cinclus*), Trauerschnapper (*Ficedula hypoleuca*), Kohl- und Tannenmeise (*Parus major*, *P. ater*). *J. Orn.* 126, 73–84.
ZINK, G. 1981. Der Zug europäischer Singvogel. *Ein Atlas der Wiederfunde beringter Vögel. Wasseramsel — Cinclus cinclus*. Lief. Vogelzug-Verlag, Moggingen.

Index